Atlas of metamorphic rocks and their textures

Atlas of metamorphic rocks and their textures

B.W.D. Yardley, W.S. MacKenzie and C. Guilford

Copublished in the United States with
John Wiley & Sons, Inc., New York

Longman Scientific & Technical.
Longman Group UK Limited,
Longman House, Burnt Mill, Harlow,
Essex CM20 2JE, England
and Associated Companies throughout the world.

Copublished in the United States with
John Wiley & Sons, Inc., 605 Third Avenue, New York, NY 10158

First published 1990

British Library Cataloguing in Publication Data

Yardley, B.W.D. (Bruce W.D.)
Atlas of metamorphic rocks and their textures.
1. Metamorphic rocks
I. Title II. MacKenzie, W.S. (William Scott) III. Guilford, C.
552′.4

ISBN 0–582–30166–1

Library of Congress Cataloging-in-Publication Data

Yardley, B.W.D.
Atlas of metamorphic rocks and their texture / B.W.D. Yardley, W.S. MacKenzie, and C. Guilford.
p. cm.
Includes bibliographical references.
ISBN 0–470–21677–8 (Wiley)
1. Rocks, Metamorphic – Atlases. 2. Petrofabric analysis – Atlases.
I. MacKenzie, W.S. II. Guilford, C. III. Title.
QE475.A2Y35 1990 90–33566
552′.4 – dc20 CIP

Set in 9/10 Times
Printed in Great Britain
by Wm Clowes, Beccles

Contents

Part 2 Textures of metamorphic rocks

Preface

The study of rocks in thin section using a petrographic microscope is an essential part of any undergraduate course in geology. This is the fourth volume in a series of photographic atlases of minerals and rocks in thin section. As in previous volumes the main purpose of the book is to provide the student with a handbook for use in practical classes to enable him or her to become familiar with the more common mineral associations and textures to be found in metamorphic rocks. In addition, some more unusual rocks which have given rise to particular significant ideas about metamorphism are also illustrated; the aim of this atlas is however to complement, not replace, a theoretical course in metamorphism.

The book has been divided into two parts. Part 1 consists of descriptions of photographs of thin sections of a wide range of rocks of different chemistries metamorphosed under a variety of physical conditions. Part 2 deals with the textures characteristic of metamorphic rocks. It is beyond the scope of this atlas to consider the origins of the rocks or to try to interpret the significance of their occurrence or their texture in depth, but in the matter of arrangement both of the rock types and their texture we have had to make some assumptions about their origins. For example in considering the pelitic rocks, we have subdivided them under the following headings: (1) medium pressure (also known as Barrovian type), (2) high temperature at medium pressure, (3) low pressure and (4) high pressure. These headings are somewhat akin to Miyashiro's different facies series.

Early studies of metamorphism considered only two main types *viz* regional metamorphism and contact metamorphism. As knowledge of the subject has increased it has become necessary to consider a greater variety of processes causing a change in mineralogy or chemistry to pre-existing rocks and, in the first chapter, examples of rocks produced by different types of metamorphism are illustrated. In general, the way in which material is presented corresponds to that adopted by Yardley (1989) *An Introduction to Metamorphic Petrology* (Longman). Some additional rock types are however illustrated here.

As in previous atlases we have tried to describe where the essential minerals appear in the photographs without the use of arrows or overprinting. We have tended to ignore individual minerals or textures which cannot be clearly seen on our original photographs because these will be even less visible in printed reproductions and there is nothing more frustrating than a photograph which does not show what it purports to show.

One or two reviews of previous atlases have noted the lack of complete petrographic descriptions of any of the rocks. This omission is intentional since we have set out to describe only what can be seen in the photographs rather than what could be seen if a thin section was available for study. This is one of the obvious limitations of a book of photomicrographs and in some cases we have tried to lessen this drawback by illustrating the rock at more than one magnification.

The number of minerals with which the student should be familiar in order to name a metamorphic rock is more than is required to give a name to the average crystalline igneous rock, but it is still a relatively small number of minerals. We have not commented on the optical properties of these common minerals except where it is useful to identify them in the photomicrographs. The relative simplicity of the nomenclature of the metamorphic rocks compared with igneous rocks is some compensation for the greater variety of minerals in the former.

A question which has to be kept constantly in mind, particularly in the study of metamorphic rocks, is how representative of the rock is the thin section? A hand specimen of the type made famous by Krantz, and found in rock collections throughout the world, measures perhaps 9 cm × 6 cm × 3 cm and thus has a volume of about 162 cubic cm: a thin section has an area of about 7 sq cm and a thickness of 0.003 cm i.e. 0.021 cubic cm in volume, and so is thus approximately one eight thousandth part of the hand specimen. In a fine-grained homogeneous rock this may be an acceptable sample but in a well foliated rock, particularly if coarse-grained, more than one thin section, cut in different orientations would be necessary to begin to describe the rock; we do not always remember to do this.

Finally we must emphasize that there is no substitute for the actual study of thin sections under the microscope. We hope however that such study can be made more rewarding for the student if he or she can, while using the microscope, compare mineral assemblages and textures with those which we have illustrated here. Although no two rocks are identical it is surprising how similar they can be, both in mineralogy and in texture, and the recognition of the same mineral assemblages appearing regularly is an indication that equilibrium is being approached. The study of textures in metamorphic rocks has given valuable insights into metamorphic processes.

Acknowledgements

We are again much indebted to colleagues and friends who have looked out thin sections for us and permitted us to take photographs of them. They include the following: S O Agrell, S Banno, K Brastad, P Brimblecombe, K Brodie, W D Carlson, D A Carswell, C Chopin, R A Cliff, G T R Droop, B W Evans, B R Frost, B Goffé, W L Griffin, S L Harley, T Hirojima, R A Howie, C B Long, I R MacKenzie, M B Mörk, J L Rosenfeld, D C Rubie, W Schreyer, J Treagus.

We are especially grateful to Dr Giles Droop of Manchester University Geology Department for looking through all of our photographs and helping us to decide which to reject in order to improve the balance between different rock types, as well as supplying us with specimens; we alone are however responsible for deficiencies in this respect.

The staff of the publishers have been very patient with us during the preparation of this book, especially as several years have elapsed since the previous atlas appeared. We hope that our experience over the intervening years in choosing suitable materials and selecting the best photographs has resulted in a better product than it might otherwise have been. Finally we wish to express our appreciation of the help given to us by Miss Patricia Crook in her accurate production of a typescript from our manuscript.

Introduction

The aim of this book is to illustrate a range of the most common and most significant metamorphic rock types, and to demonstrate the way in which deductions can be made about metamorphic conditions and the metamorphic history of a region, from observations in thin section.

Metamorphism occurs as a response to changes in the physical or chemical environment of any pre-existing rock, such as variations in pressure or temperature, strain, or the infiltration of fluids. It involves recrystallization of existing minerals into new grains and/or the appearance of new mineral phases and breakdown of others. Metamorphic processes take place essentially in the solid state, in that the rock mass does not normally disaggregate and lose coherence entirely, however small amounts of fluids are frequently present and may play an important catalytic role; at very high grades, melts may be produced.

Metamorphic settings

In this book we have followed the classification of metamorphic settings used by Yardley (1989):

Contact metamorphism takes place as a result of heating of the country rocks in the immediate vicinity of igneous intrusions or beneath thick flows. It is characterized by the growth of new metamorphic mineral grains in random orientations, since any deformation is usually too weak to produce marked mineral alignments. Contact metamorphism is also known as thermal metamorphism, and its typical products are rocks known as hornfels.

Regional metamorphism gives rise to large tracts of metamorphic rocks characteristic of many mountain belts and ancient shield areas. Typically regional metamorphism involves heating, burial to produce elevated pressures controlled by the depth attained in the crust or mantle, and deformation to produce tectonic fabrics. *Burial metamorphism* is a form of regional scale metamorphism that takes place at low temperatures ($<$ circa 250 °C) in the absence of appreciable deformation.

Dynamic metamorphism occurs in response to intense strain and hence is usually of localized occurrence, notably in shear zones.

Hydrothermal metamorphism involves chemical reactions brought about by circulating fluids and is often accompanied by a change in the chemical composition of the rock, known as *metasomatism*. The most widespread occurrence of hydrothermal metamorphism is *sea-floor metamorphism* taking place at active spreading centres. In contrast, much metamorphism involves little chemical change except loss of volatiles, and is termed *isochemical*.

Impact metamorphism has no genetic relation to the other types and is brought about by the impact of large, high-velocity meteorites on planetary surfaces. It results from extreme shock effects and can produce dense minerals, normally formed only at mantle depths, on the earth's surface.

With the exception of the last, these categories are not entirely distinct. Instead they grade into one another as a result of different processes acting together; for example, intense strain can occur locally within a region undergoing regional metamorphism. It is also possible for rocks within a broadly regional metamorphic belt to have been subjected to different types of metamorphism at different times in their history.

Metamorphic rock names

The terminology of metamorphic rock names used here is that of Yardley (1989), from which the following passage is taken:

> There are four main criteria for naming metamorphic rocks:
> 1. the nature of the parent material;
> 2. the metamorphic mineralogy;
> 3. the rock's texture;
> 4. any appropriate special name.

Names indicating the nature of the parent material

These may be very general e.g. metasediment, or more specific e.g. marble. Such names may be used as nouns with or without additional qualification e.g. diopside marble, or as adjectives qualifying a textural name e.g. pelitic schist. Some of the common names, and their adjectival forms are as follows:

Original material	*Metamorphic rock type (noun/adjective)**
Argillaceous or clay-rich sediment	Pelite/pelitic
Arenaceous or sandy sediment	Psammite/psammitic or quartzofeldspathic (if appropriate)
Clay–sand mixture	Semi-pelite
Quartz sand	Quartzite
Marl	Calc-silicate/calcareous
Limestone	Marble
Basalt	Metabasite/mafic
Ironstone	Meta-ironstone/ ferruginous

* In addition, it is acceptable to prefix any igneous or sedimentary rock name by *meta-* to denote the metamorphic equivalent, as in the last two examples.

Metamorphic mineralogy

The names of particularly significant metamorphic minerals that may be present are often used as qualifiers in the metamorphic rock names, e.g. garnet mica schist, forsterite marble. There are two possible conventions here: the mineral names may be given in order of abundance for the principal metamorphic minerals, to denote the modal mineralogy, e.g. garnet sillimanite schist; or the names of particularly significant minerals can be given, which indicate specific conditions of metamorphism, irrespective of their abundance, e.g. sillimanite muscovite schist. The first convention might be more appropriate for a field geologist who wishes to make stratigraphic correlations and can use the modal mineralogy as a rough guide to rock composition. On the other hand a petrologist studying variations in metamorphic grade will specify only those minerals that indicate particular conditions of metamorphism. Some essentially monomineralic rocks are named for their dominant mineral e.g. quartzite, serpentinite or hornblendite. A number of other names referring to particular mineral associations are described under *Special names* below.

The rock's texture

Textural terms are very important for naming metamorphic rocks and indicate whether or not oriented fabric elements are present to dominate the rock's appearance, and the scale on which they are developed. Although mineral preferred orientations are best developed in pelites and semi-pelites, they can form in a wide range of rock types if deformation is sufficiently intense. In many regionally metamorphosed rocks micas develop a preferred orientation, aligned perpendicular to the maximum compression direction, giving rise to a planar fabric or foliation. The names used for planar fabrics depend on the grain size and gross appearance of the rock. Deformation and metamorphism of clay-bearing clastic sediments give rise to the following sequence of rocks with characteristic fabrics, in order of increasing grade of metamorphism:

Slate – a strongly cleaved rock in which the cleavage planes are pervasively developed throughout the rock, due to orientation of very fine phyllosilicate grains. The individual aligned grains are too small to be seen with the naked eye, and the rock has a dull appearance on fresh surfaces.

Phyllite – similar to slate but the slightly coarser phyllosilicate grains are sometimes discernible in hand specimen and give a silky appearance to cleaved surfaces. Often, the cleavage surfaces are less perfectly planar than in slates.

Schist – characterized by parallel alignment of moderately coarse grains usually clearly visible with the naked eye. This type of fabric is known as schistosity and where deformation is fairly intense it may be developed by other minerals, such as hornblende, as well as by phyllosilicates.

Gneiss – gneisses are coarse, with a grain size of several millimetres, and foliated (i.e. with some sort of planar fabric, such as schistosity or compositional layering). English and North American usage emphasizes a tendency for different minerals to segregate into layers parallel to the schistosity, known as gneissic layering; typically quartz- and feldspar-rich layers segregate out from more micaceous or mafic layers. European usage of gneiss is for coarse, mica-poor, high grade rocks, irrespective of fabric. The term *orthogneiss* is used for gneisses of igneous parentage, *paragneiss* for metasedimentary gneisses.

In practice the boundaries between all these types are gradational.

Mylonite – is a term used for fine-grained rocks produced in zones of intense ductile deformation where pre-existing grains have been deformed and recrystallized as finer grains.

Hornfels – contact metamorphism in the absence of deformation gives rise to a random fabric of interlocking grains which produces a tough rock known as *hornfels*.

Some metamorphic rocks, particularly those relatively poor in sheet silicates, have textures that are not obviously schistose, even though the rocks are not hornfelses. Winkler (1976) has proposed the term *fels* for such rocks, although it has not been universally adopted. In the older literature the term *granulite* is used for some such rocks, particularly psammites with an equigranular texture, but this term is now reserved to denote particular physical conditions of metamorphism.

Textural names are usually used as nouns, qualified by adjectives indicating the parent material or the present mineralogy (e.g. garnet schist, pelitic hornfels).

Special names

Special names are mercifully rare in metamorphic petrology and most that are used are also descriptive. However, the mineral associations indicated by the names carry implications for the conditions of metamorphism. Some of the commonest are the following:

Greenschist – green foliated metabasite, usually composed predominantly of chlorite, epidote and actinolite.

Blueschist – dark, lilac-grey foliated metabasite, owing its colour to the presence of abundant sodic amphibole, typically glaucophane or crossite: seldom truly *blue* in hand specimen.

Amphibolite – an essentially bimineralic dark green rock made up of hornblende and plagioclase. A wide range of minerals may occur as accessories. Most amphibolites are metabasites (ortho-amphibolite) but some may be metamorphosed calcareous sediments (para-amphibolites).

Serpentinite – green, black or reddish rock composed predominantly of serpentine. Formed by hydration of igneous or metamorphic peridotites (olivine-rich ultrabasic rocks).

Eclogite – metabasite composed of garnet and omphacitic clinopyroxene with no plagioclase feldspar. Common accessories include quartz, kyanite, amphiboles, zoisite, rutile or minor sulphides.

Granulite – rock characterized by both a texture of more or less equidimensional, straight sided (polygonal) grains for all mineral species, and a mineralogy indicative of very high temperature metamorphism, closely akin to the mineralogy of calc-alkaline basic to moderately acid plutonic rocks (feldspar, pyroxene, amphibole). The *charnockite* suite constitutes a distinct variety of K-feldspar and hypersthene bearing granulites.

Migmatite – a *mixed rock* composed of a schistose or gneissose portion intimately mixed with veins of apparently igneous quartzo-feldspathic material (known as leucosomes).

Textural terms

The textural terms used in the descriptions in this volume are introduced at the beginning of Part 2.

Physical conditions of metamorphism: metamorphic facies

One of the most important goals of metamorphic petrology is to determine the pressures (P) (and hence depths) and temperatures (T) at which particular rocks formed. An account of this is entirely beyond the scope of this book and we deal here only with those aspects which are essential to understanding the layout of the main part of the book.

The appearance of particular metamorphic minerals depends on both the bulk composition of the rock and the P–T conditions which it experienced. With increased heating for example, pelitic schists develop a sequence of progressively higher temperature assemblages of minerals. The metamorphic terrane can therefore be divided into zones each characterized by a particular mineral or suite of minerals. Rocks subjected to higher temperatures and pressure are said to be higher *grade* than those subjected to less extreme conditions. Zone boundaries represent a constant grade and so are known as *isograds*.

Different original rock types respond in different ways to the same conditions of metamorphism according to their bulk composition, and some show far fewer mineralogical changes than others. For this reason it is not usually possible to trace zones defined by assemblages in one rock type through regions where that rock type is absent. To overcome this problem Eskola (1915) devised a scheme of more broadly defined *metamorphic facies* each corresponding to regions of the *P–T* diagram which may be distinguished by the assemblages in any one of a range of rock types. The assemblages of metabasites are however the primary basis for the facies classification.

The scheme of metamorphic facies used here is illustrated in Figure A.

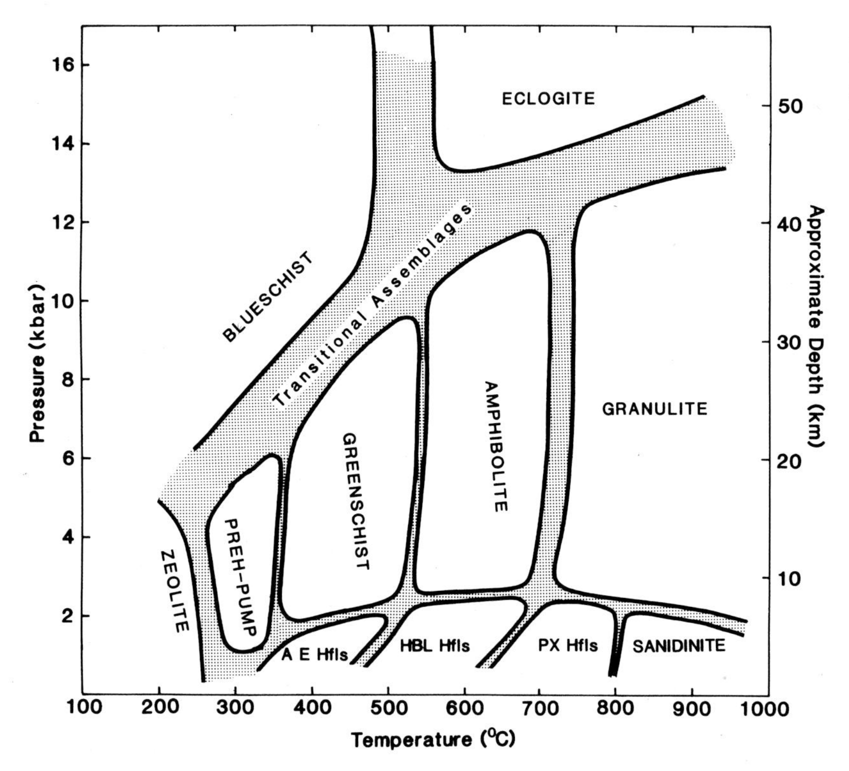

FIG. A Pressure–temperature diagram showing the fields of the various metamorphic facies, after Yardley (1989). Hfls = hornfels, AE = albite–epidote, HBL = hornblende, PX = pyroxene, PREH–PUMP = prehnite–pumpellyite

Using this book

We have divided this volume into two parts: in the first the primary aim is to illustrate some important metamorphic mineral assemblages grouped according to parental rock composition and *P–T* conditions of metamorphism; Part 2 specifically illustrates textures.

Part 1 is divided into sections according to parental rock type, corresponding in part to chapters in Yardley (1989). Sequences of photomicrographs first illustrate progressive metamorphic zones encountered in normal medium pressure metamorphism, followed by examples of unusually high temperature metamorphism at intermediate pressures. After this, as appropriate, are examples of lower pressure metamorphic sequences and higher pressure sequences.

Part 2 illustrates some basic textural terminology and then has sections illustrating deformational textures, reaction textures and timing relationships of deformation and porphyroblast growth. Inevitably there is considerable overlap between the two parts and we have provided cross referencing as appropriate. Thus, additional examples of many of the pelite assemblages from Part 1 are found in Part 2.

In writing the brief descriptions to accompany the photographs we have assumed that the reader has some general familiarity with optical mineralogy, but we have provided sufficient details of the properties of the more unusual minerals to enable them to be identified. Where a mineral's colour is given, this refers to its colour in plane polarized light unless interference (or birefringence) colour is specified. The abbreviations PPL and XPL are used for plane polarized light and cross polarized light respectively. Where it is necessary to indicate the location of particular features this is sometimes done using notional geographic coordinates: north (N) is the top of the photograph, etc. In addition, N–S features may be referred to as vertical, and so on. Cross references in **bold** numerals refer to the rock number; more rarely, where cross references are to page numbers, this is specified. For some rocks, a specific literature reference to a paper describing them is given at the end of the caption. General references given elsewhere in the text are listed at the end of the book.

Part 1

Varieties of metamorphic rocks

Contact metamorphism

Contact or *thermal metamorphism* occurs in the country rocks surrounding an igneous intrusion as a result of magmatic heating. The resulting metamorphic rocks comprise a metamorphic aureole around the intrusion or group of intrusions that provided the heat source, and often show concentric metamorphic zones.

Most typically, contact metamorphism results in the production of *hornfelses* i.e. rocks whose metamorphic minerals have intergrown in a random interlocking pattern because of the absence of strain as they grew, however in some aureoles magma emplacement was accompanied by deformation of the country rocks, leading to the formation of contact metamorphic schists texturally similar to those produced in regional metamorphism. Contact metamorphism can affect a wide range of rock types, but most aureoles are developed in metasedimentary rock with a previous history of regional metamorphism, usually at low to moderate grades. However contact metamorphism of sediments is common at high, subvolcanic levels, and aureole effects in pre-existing igneous rocks occur also.

In this section we illustrate two classic examples of hornfels. The cordierite chlorite biotite hornfels (**1**) is typical of *spotted slates*, produced by thermal metamorphism of slates around granitic plutons, while the peridotite hornfels (**2**) is a more unusual rock type which displays very graphically the way in which newly-formed metamorphic minerals can grow to produce an interlocking texture.

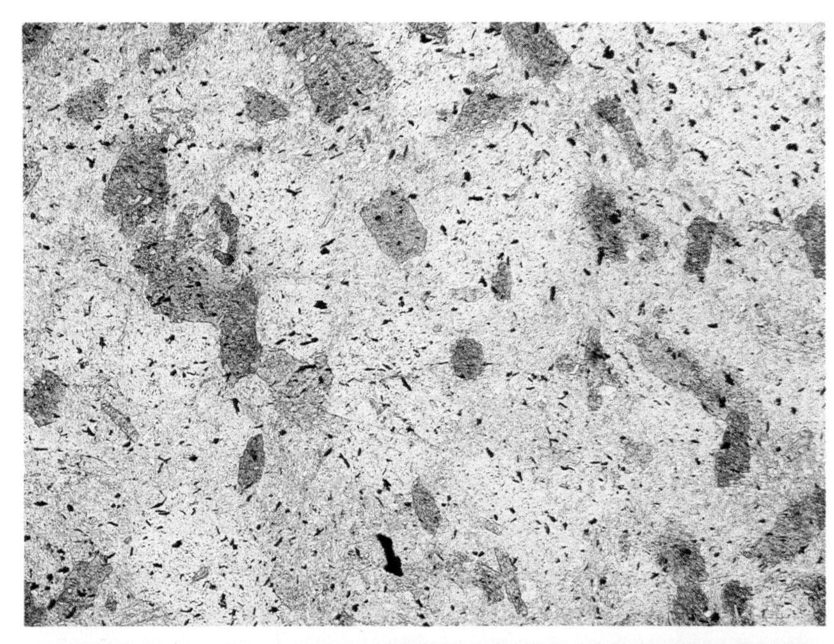

1

Cordierite chlorite biotite hornfels

Contact metamorphism

This rock shows elongate crystals of brown biotite and finer grained, green, low birefringence chlorite in random orientation typical of a hornfels. The main colourless mineral in the rock is poikiloblastic cordierite, recognized readily from the XPL view in which cyclic (sector) twinning is seen. The matrix between the porphyroblasts comprises a fine-grained intergrowth of muscovite, opaque grains and quartz.

Locality: Skiddaw aureole, England. Magnification: × 52, PPL and XPL.

2

Peridotite hornfels

Contact metamorphism

The characteristic hornfels texture of randomly oriented interlocking crystals is particularly well displayed by this rock, although its composition is unusual for a hornfels. It is composed predominantly of olivine and orthopyroxene, with the orthopyroxene forming randomly oriented prisms showing cleavage and relatively low birefringence, set in an olivine matrix. Highly birefringent material replacing some orthopyroxene grains is talc.

This rock occurs in the aureole of a major batholith, where it cuts a serpentinite body. The heat from the intrusion has broken down serpentine and more or less restored the original igneous mineralogy of the ultramafic rock, albeit with a distinctive texture.

Locality: Mount Stuart, Northern Cascades, Washington, USA. Magnification: × 14, PPL and XPL.

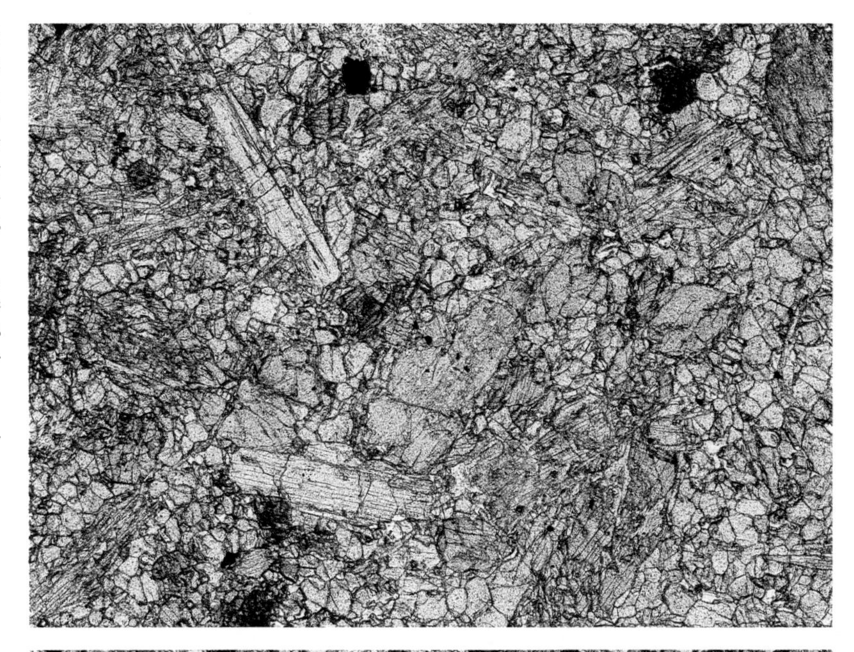

Regional metamorphism

Regional metamorphism is usually more extensive than contact metamorphism and is not closely focused around a specific magmatic heat source; indeed no heat source may be apparent. Typically, the growth of new metamorphic minerals in regional metamorphism is accompanied by deformation and by the production of tectonic mineral fabrics in response to strain.

Textural studies show that although metamorphic mineral growth broadly accompanies deformation in regional metamorphism, in detail different deformation episodes may have occurred and mineral growth does not always correspond with periods of deformation (p. 94).

Most metamorphic rocks have undergone predominantly regional metamorphism, and there is a very wide range of conditions of pressure and temperature over which regional metamorphism can occur. At the high temperature, low pressure end of the spectrum of metamorphic facies, regional metamorphism is usually closely associated with the emplacement of magmas; there is no fundamental division between (*i*) regional metamorphism driven by magmatic heating from multiple intrusions, so that there is no single focus, and (*ii*) contact metamorphism at similar pressures and temperatures localized in an aureole to a specific intrusion. Regional metamorphism may also overprint earlier hydrothermal metamorphism, notably in metavolcanic rocks.

Regional metamorphic rocks often contain zones of high strain, especially along shear zones and faults, within which the rock texture is dominated by deformational effects. In this case, regional metamorphism becomes transitional to dynamic metamorphism.

Dynamic metamorphism

Dynamic metamorphism is dominated by deformation and recrystallization due to strain, and is usually accompanied by a reduction in grain size. The name *mylonite* is used for rocks that have undergone dynamic metamorphism, and mylonites are usually of restricted occurrence within fault zones, including thrusts and shears. However some shear zones can be several kilometres in width and extend for tens or even hundreds of kilometres.

Dynamic metamorphism is a process that progressively affects pre-existing metamorphic or igneous rocks, only destroying all trace of original fabrics if it is very intense. Since ductile deformation processes are involved, temperatures are likely to be in excess of about 300 °C, and so truly unmetamorphosed sediments are unlikely to be affected.

Different minerals respond in very different ways to deformation. In crustal rocks containing quartz, it is the quartz that deforms most readily, forming strained grains with undulose extinction, which then break down to a finer-grained matrix of undeformed grains through the process of syntectonic recrystallization. Minerals such as feldspar and garnet are relatively strong, and often remain as relatively large relic grains, somewhat rounded by breaking-off or recrystallization of their edges and corners. These grains are known as porphyroclasts. Micas and other phyllosilicates readily recrystallize in mylonites and may also be produced by hydration reactions due to infiltration of water into the zone of deformation.

A range of siliceous mylonites are illustrated later in the book (**84–88**, pp. 90–94), the example shown here is more unusual, being a mylonite of ultrabasic composition produced by deformation of peridotite under upper mantle conditions. At high temperatures in olivine-rich rocks it is the olivine that deforms most readily, while pyroxenes, garnet or spinel form porphyroclasts.

3
Garnet biotite schist

Regional and dynamic metamorphism

This highly deformed rock is composed of quartz and muscovite with porphyroblasts of garnet and biotite. The foliation is intense and muscovites show remarkable parallelism; in addition the rock is segregated into mica-rich and quartz-rich domains parallel to the schistosity. Much of the opaque material here is graphite. Note that the foliation tends to wrap around the porphyroblasts due to deformation after they grew. Biotite porphyroblasts are attenuated in the direction of shearing to form a distinctive shape known as 'mica fish'. Coarser quartz grains have sutured boundaries and some show strained extinction; they appear to have been arrested in the process of breaking down to finer grains during syntectonic recrystallization.

This sample was collected only a short distance from the major Alpine Fault of New Zealand.

Locality; Franz Joseph Glacier, Westland, New Zealand. Magnification: × 10, PPL and XPL.

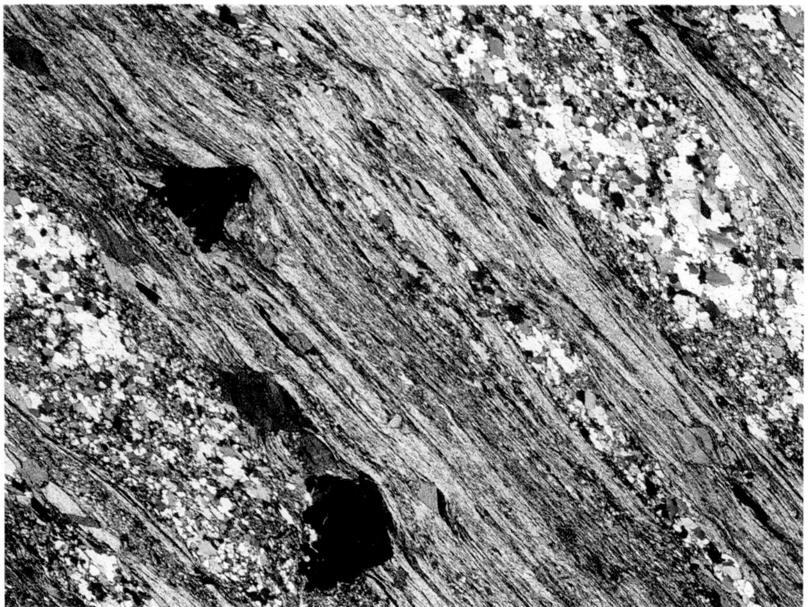

4

Garnet staurolite schist/ biotite schist

Regional metamorphism with weak deformation only

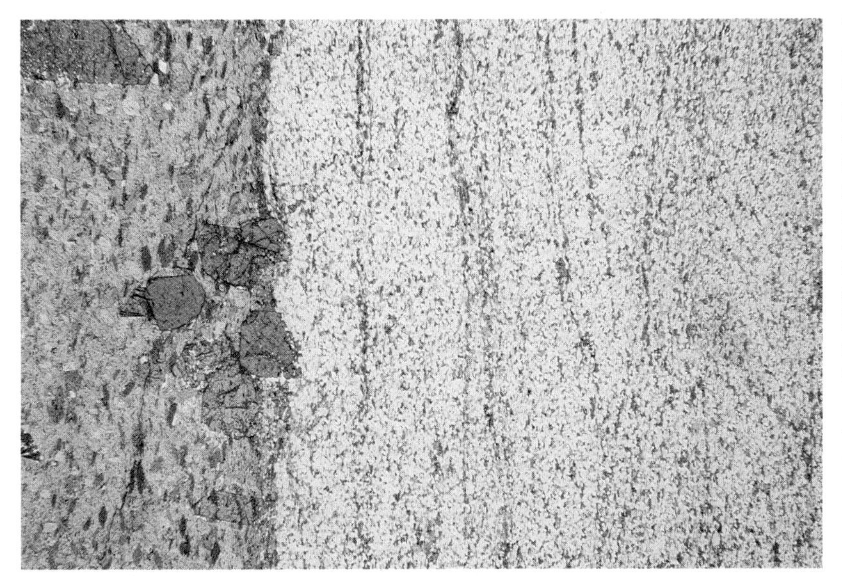

This low magnification view shows the contact between two contrasting original beds, one of clay (pelite), the other of sand (psammite). The pelite band is now composed of euhedral porphyroblasts of staurolite (high relief, pale yellow) and garnet (high relief, grey) in a matrix of biotite, muscovite and minor quartz and ilmenite. The assemblage of the pelite layer indicates amphibolite facies metamorphism, but the psammite layer nevertheless retains evidence of original sedimentary structure. The biotite-rich stripes within it mimic original bedding and pick out cross-stratification. Note that the grain size of the originally coarser psammite has probably changed relatively little, whereas the original clay band is now much coarser.

Locality: Coos Canyon, Rangely District, Maine, USA. Magnification: × 5, PPL.

5

Peridotite mylonite

Dynamic metamorphism

This rock is a protomylonite of unusual composition. Porphyroclasts are of olivine and pyroxene (both clino- and ortho-), and may have been markedly bent during deformation, so that extinction varies considerably along the length of the grains. Some porphyroclasts have long, drawn out *tails*, which break up or recrystallize to contribute to the finer-grained matrix, itself made up of the same minerals as occur as porphyroclasts. An elongate, low birefringence orthopyroxene grain near the centre of the field of view contains fine pale E–W stripes that are lamellae of clinoenstatite. This mineral is extremely rare, occurring only as the result of stress-induced polymorphic transition from enstatite.

Small isotropic grains of dark brown spinel are present and are also highly deformed.

Locality: Premosello, Val d'Ossola, Northern Italy (Ivrea Zone). Magnification: × 7, XPL.

Sea-floor and hydrothermal metamorphism

Hydrothermal metamorphism can occur in a wide range of settings but is characterized by the involvement of hot aqueous fluid which passes through the metamorphosing rock and leads to changes in its chemical composition or *metasomatism*. The extent of such changes can be rather minor, dominated by hydration, or may be extensive and result in the formation of a monomineralic metasomatic rock in which the abundances of many of the chemical elements in the rock have been changed.

Although localized hydrothermal metamorphism is common around igneous intrusions and in shear zones and faults, by far the most important occurrences volumetrically are produced by interaction of heated sea-water with newly-created oceanic crust at mid-ocean ridges. This type of hydrothermal metamorphism is known as *sea-floor metamorphism* and is found in ophiolites on land as well as on the present ocean floor. Rocks with an early history of sea-floor metamorphism may subsequently undergo regional metamorphism, and the second of the examples illustrated has probably undergone a complex metamorphic history of this type (**7**).

6

Sea-floor amphibolite

Sea-floor metamorphism

This sample was collected in a dredge haul from the ocean floor. It is a very fine-grained rock with a random texture that may mimic an original fine-grained ophitic basalt texture. The dominant constituents are pale green actinolite and plagioclase, with appreciable opaque oxide. Small amounts of calcite are also present.

Locality: Peake Deep Area, Atlantic Ocean. Magnification: × 38, PPL and XPL.

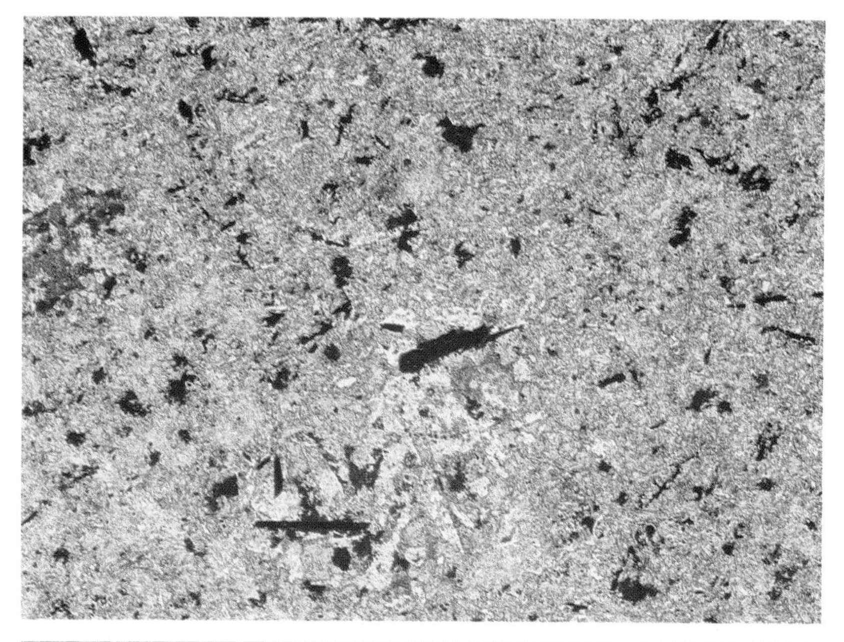

7

Epidosite

Hydrothermal metamorphism

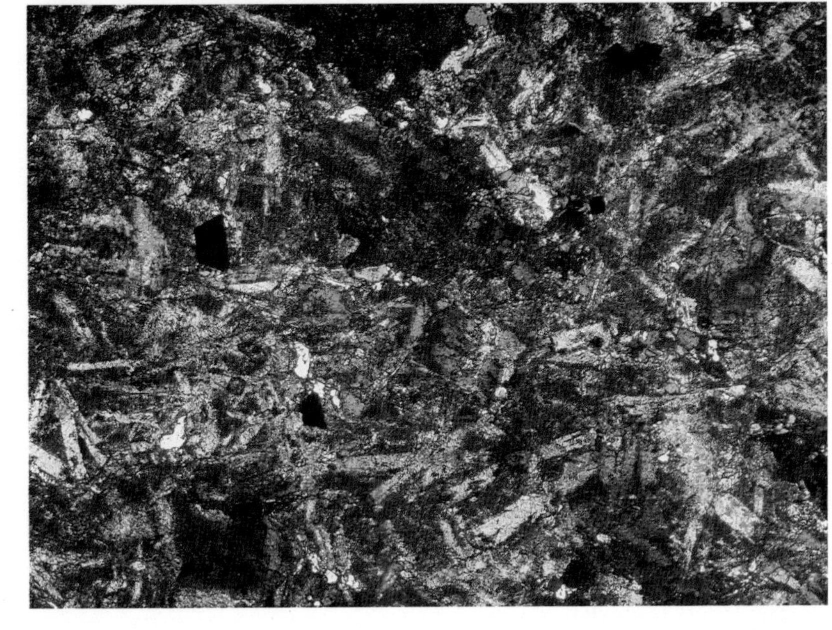

This rock is a product of hydrothermal metamorphism accompanied by extensive metasomatism. It is composed predominantly of epidote, but minor green chlorite and clear quartz crystals are visible in the PPL view. The XPL view is an enlargement of part of the area shown in PPL, and also contains minor quartz. The texture seen in PPL appears to be a typical ophitic texture of basalt. However both the clear yellow laths, corresponding to original plagioclase, and the intervening brown areas (originally pyroxene or glass) are now epidote. Careful inspection of the areas of uniform birefringence colour in the XPL view shows that single epidote crystals present now are coarser than the original grain size and have a granoblastic texture. Hence clusters of adjacent laths are now ghosts within a single epidote grain.

Locality: Claggan Bay, Achill Island, Ireland. Magnification: × 22, PPL and × 45, XPL.

Impact metamorphism

Impact metamorphism has no genetic link with the other categories of metamorphism, and affects rocks on the earth's surface close to the site of impact of large high-velocity meteorites. Such events are very rare on Earth and ancient meteorite impact sites have usually been extensively reworked by erosion and other geological processes. On tectonically inert planets like the Moon however, meteorite impact may be the dominant geological process reworking the planetary surface.

The shock wave that passes out from the point of impact subjects rocks to pressures normally only experienced at mantle depths for extremely short periods of time, while the subsequent stress relaxation leads to temperatures that may be sufficiently hot to melt or even vapourize the rock. Shock effects dissipate outwards from the site of impact and range from fracturing of rock and internal deformation of grains to the production of high pressure mineral polymorphs (such as the dense forms of SiO_2, coesite and stishovite) or melting.

8

Impact metamorphic rock

This sample illustrates a range of features typical of rocks produced by intense shock metamorphism at a meteorite impact site. The rock contains angular fragments of quartz, feldspar and biotite set in a fine matrix that is in large part glass produced by impact melting. The colour of the glass is variable because of its extreme chemical heterogeneity, whilst the crystalline material comprises angular fragments of original coarse-grained granite basement material, and so is distinct from volcanic phenocrysts. Biotite in the lower right corner has been distinctly bent by the impact event.

Locality: Ries Crater, Germany. Magnification: × 43, PPL and XPL.

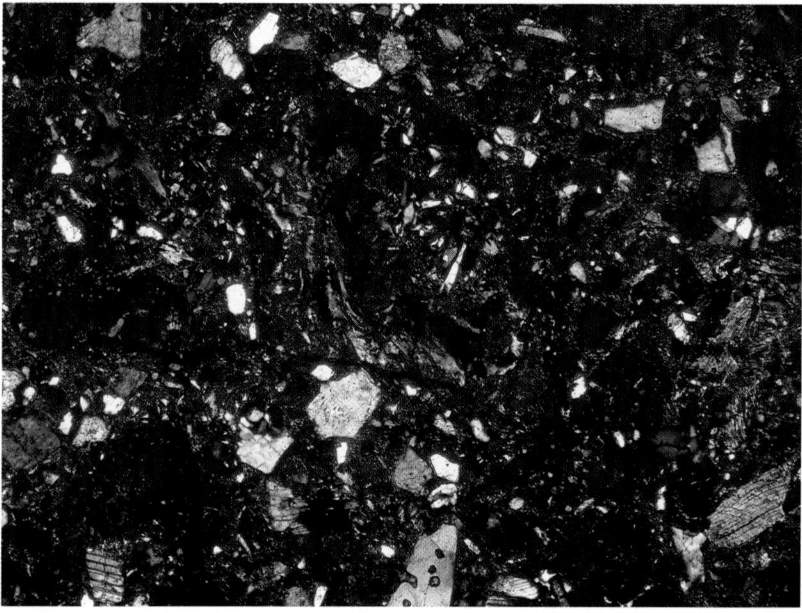

Metamorphism of pelitic rocks

Medium pressure metamorphism

The term *Barrovian metamorphism* is one that is widely used in the English-speaking world to describe medium grade metamorphism that has taken place at moderate pressures i.e. over a range of *P–T* conditions corresponding approximately to a normal geothermal gradient in continental crust. The name derives from the work of G M Barrow, in the late 19th century, on metamorphic zones in the southern Highlands of Scotland (Barrow, 1893). Barrovian metamorphism spans the temperature range of the greenschist and amphibolite facies (Figure A), at sufficiently high pressures, such that kyanite rather than andalusite is the first Al_2SiO_5 polymorph to appear on heating.

A similar pattern of metamorphism to that found by Barrow has been reported from many parts of the world and the pelite zones are illustrated in the following section in order of increasing metamorphic grade. They include examples of both lower and higher grade medium pressure metamorphism than those occurring in Barrow's type area.

9

Graphitic slate

Prehnite pumpellyite facies

This very fine-grained rock represents the lowest grade of metamorphism. At an advanced stage of diagenesis, the clay minerals are dominantly chlorite and illite, and with further metamorphism the illite itself becomes coarser-grained and recrystallizes to a phengitic mica which is richer in Si and poorer in Al than pure muscovite and contains some Mg and Fe.

This rock contains detrital grains of quartz and minor alkali feldspar with a fine matrix of phengitic mica, graphite and some chlorite.

The rock has been intensely deformed producing a pervasive slaty cleavage and at the same time, original fine scale bedding has been disrupted by folding. Fragmented silty layers rich in detrital quartz appear as lighter regions in a darker pelitic matrix. Note that the slaty cleavage cuts across contacts between the bed types and is itself cut by two late-stage veinlets.

Locality: Routeburn Track, South Island, New Zealand.
Magnification: × 12, PPL.

10

Chlorite muscovite albite schist

Greenschist facies – chlorite zone
(Additional example: **83**)

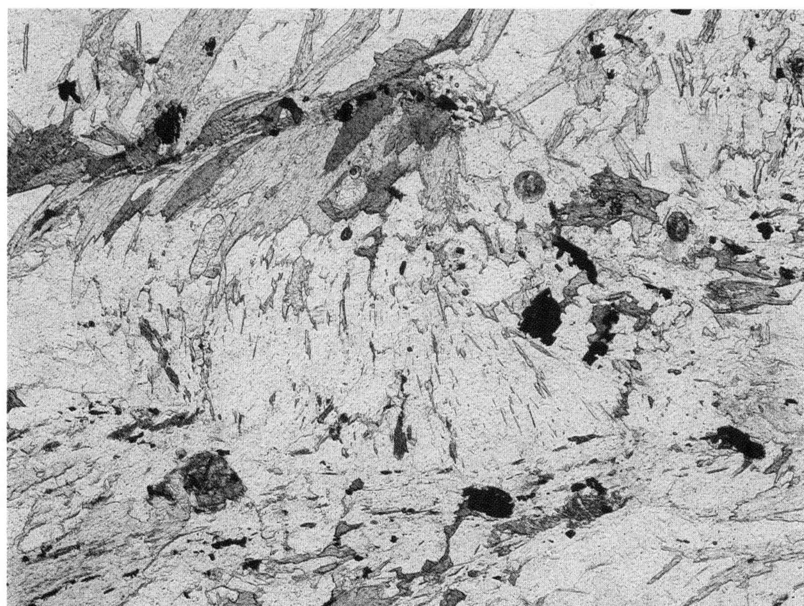

This rock is from the chlorite zone of the Dalradian of the British Isles. The chlorite–muscovite intergrowths can be well seen in the higher magnification photographs; the pale green colouration of the muscovite is the result of an appreciable phengite content. The colourless minerals are quartz and albite, the latter forming distinct porphyroblasts sometimes simply twinned as illustrated here. Accessory minerals visible in the high magnification view include apatite, occurring as colourless, near-isotropic high relief grains enclosed in muscovite and albite, opaque oxides, and a small zircon (within the albite at its upper edge). Flaws in the section appear as circular high relief areas in the upper right quadrant of the × 30 views.

This rock displays a pronounced crenulation fabric, with an earlier pervasive phyllosilicate fabric folded to produce a new spaced cleavage. The higher magnification view shows that the albite porphyroblasts overgrow both the fabrics, and therefore grew post-tectonically.

Locality: Cloghmore, southeast Achill Island, Ireland. Magnification: × 14, XPL; and × 30, PPL and XPL.

11

Biotite chlorite muscovite schist

Greenschist facies – biotite zone
(Additional example: **92**)

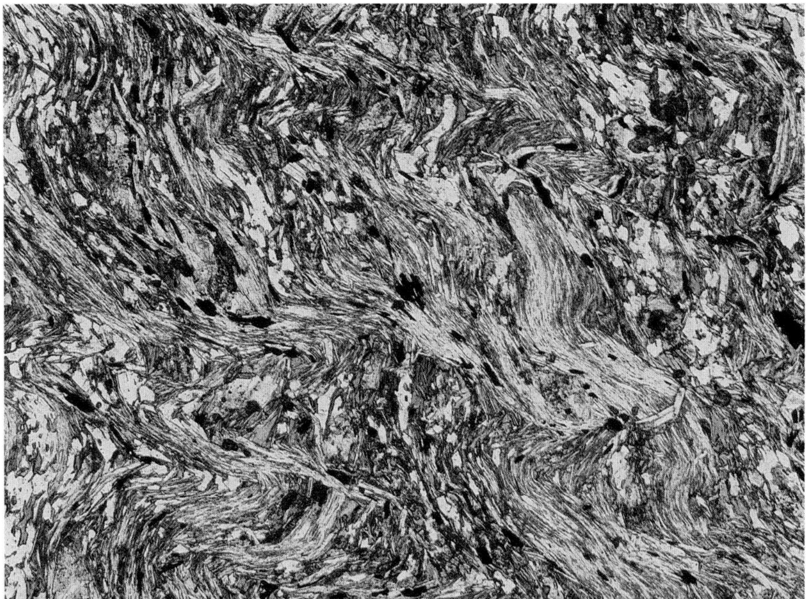

The bright interference colours in this rock are mainly due to the high proportion of muscovite which is present. The biotite and chlorite can be seen readily in the PPL view. Other minerals present are dominated by quartz; albite (now partly sericitized) has a patchy pale brown appearance in PPL and there is a small percentage of opaque grains.

The small scale folding of the original schistosity in which the platy minerals were aligned has produced a crenulation cleavage (*see* texture section). This has been accompanied by some segregation of the quartz into horizontal layers corresponding to crenulation hinges and separated by layers that are nearly pure phyllosilicate.

Locality: northwest Mayo, Ireland. Magnification: × 27, PPL and XPL.

12
Microcline epidote mica schist

Greenschist facies – biotite zone

This is a semi-pelitic rock consisting of green biotite, muscovite, epidote, microcline and quartz. Reaction between chlorite and microcline produces biotite at a slightly lower grade than in pelitic rocks lacking K-feldspar, and this reaction accounts for the absence of chlorite in this rock.

The high RI mineral showing bright interference colours is epidote; a small grain is present next to the upper edge, half-way along it.

Locality: northwest Mayo, Ireland. Magnification: × 20, PPL and XPL.

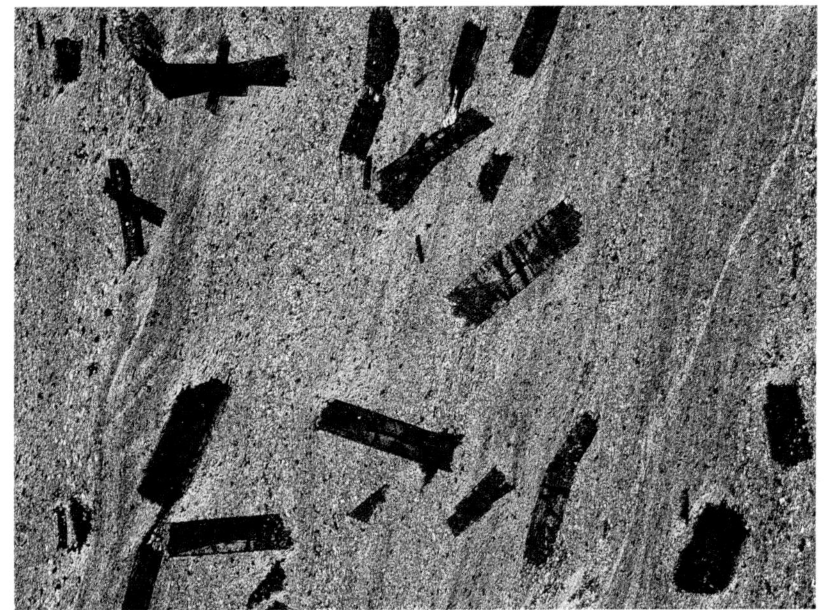

13
Chloritoid slate

Greenschist facies – biotite zone

A fine-grained slate containing randomly oriented chloritoid porphyroblasts, which in this specimen are manganese-rich and have the variety name ottrelite. The fine-grained groundmass of the rock consists of chlorite, muscovite, quartz and hematite. Note that fine scale sedimentary layering is well preserved, cut by an oblique slaty cleavage, despite the fact that the chloritoid is of comparable size to the spacing of the original laminations. Some of the chloritoid crystals are so full of inclusions that they appear almost opaque. The hour glass structure is not uncommon in chloritoid.

Locality: South of Vielsalm station, Ardennes, Belgium. Magnification: × 20, XPL.

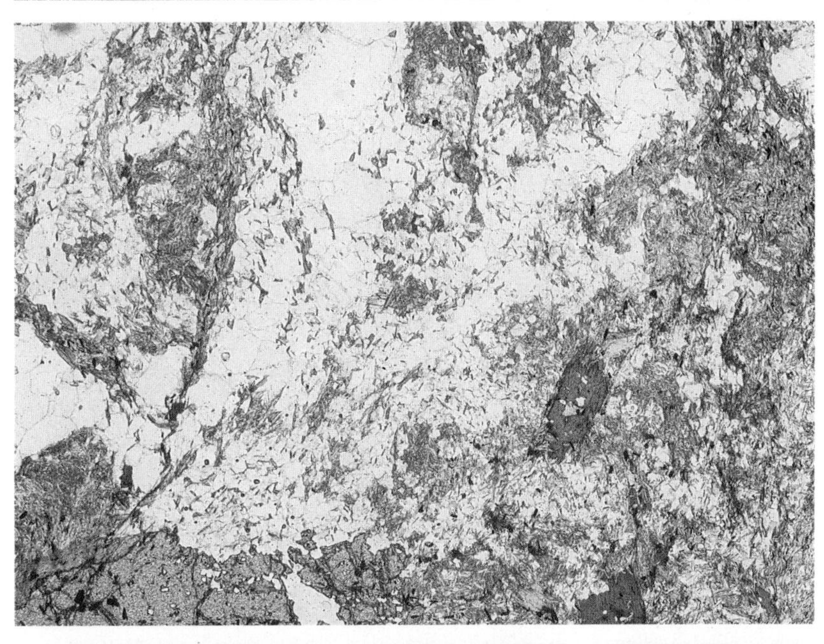

14
Garnet chlorite biotite schist

Greenschist facies – garnet zone
(Additional examples: **82, 89, 91, 99**)

The diagnostic garnet zone pelite assemblage of garnet + biotite + chlorite + muscovite + quartz is well displayed in this sample. The rock has a markedly porphyroblastic texture with very large (<1 cm) euhedral or subhedral garnet grains in a fine matrix. Biotite is also porphyroblastic though much finer-grained than garnet.

Chlorite and muscovite occur with quartz in the matrix and define a complex fabric formed over at least two stages of deformation which apparently pre-dated the peak metamorphic temperatures at which biotite and garnet grew.

Locality: Bridgewater Corners, Vermont, USA. Magnification: × 18, PPL and XPL.

15

Garnet chloritoid schist

Greenschist facies – garnet zone

The mineral assemblage seen here is typical of highly aluminous pelites in the garnet zone of Barrovian type metamorphism. Note that biotite is absent in most such chloritoid schists.

Chloritoid is recognized by its green absorption colours and high relief. Different grains show three distinct colours, and the palest, straw colour is very similar to that of garnet. This is illustrated by the two PPL views, taken with the polarizer at right angles. The much lower RI of the chlorite makes it possible to distinguish readily between the two green minerals. Only one garnet is shown, just below the centre of the field. The other minerals present are muscovite, quartz and albite.

Locality: Ebeneck, 6 km northwest of Mallnitz, Kärnten, Austria. Magnification: × 22, PPL, PPL and XPL.

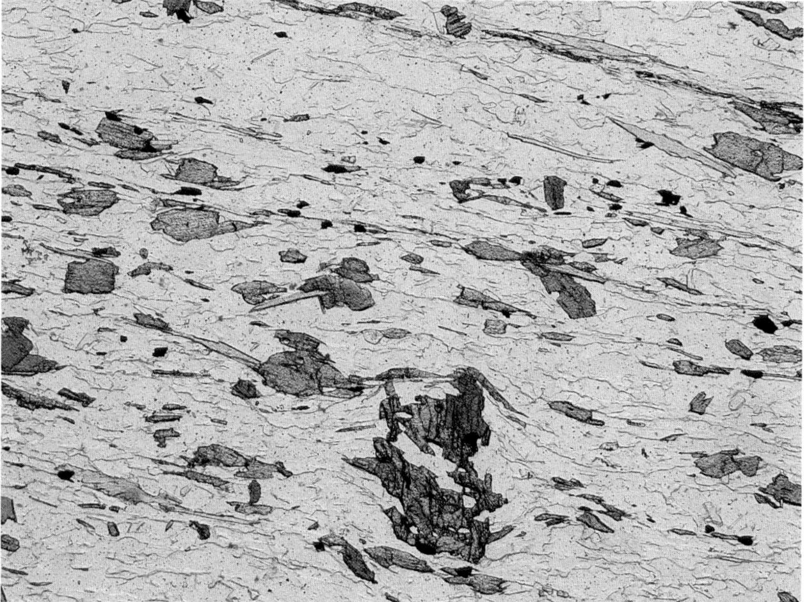

16

Staurolite schist

Amphibolite facies – staurolite zone
(Additional examples: **90, 94, 95, 97**)

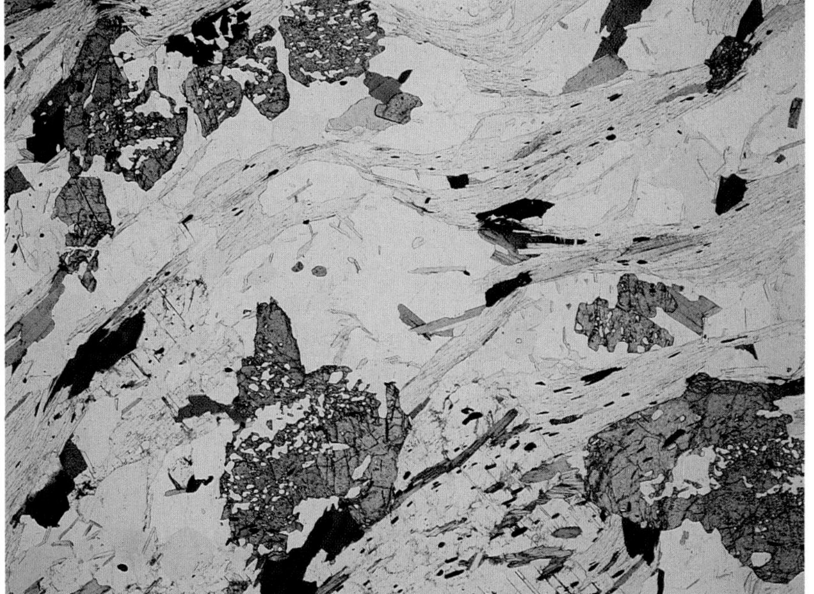

This rock contains poikiloblasts of high relief staurolite which display more marked yellow pleochroism than is usual. The largest poikiloblasts are of plagioclase (e.g. near mid-point of lower edge); other minerals present are muscovite, quartz, green biotite and an opaque mineral. The section is slightly thick and in consequence quartz crystals have a yellowish tinge to their interference colour.

The schistosity in this rock is defined by both muscovite and the opaque grains, and is seen in the lower right corner to pass continuously into the inclusion trails within plagioclase, without disruption. Note that the quartz inclusions within staurolite are very fine, whereas the matrix quartz is very coarse, evidently there was extensive recrystallization of quartz after staurolite grew.

Locality: Connecticut, USA (precise locality unknown). Magnification: × 7, PPL and XPL.

17

Garnet staurolite kyanite gneiss

Amphibolite facies – kyanite zone

The upper part of the larger field of view is mainly occupied by porphyroblasts consisting of epitaxial intergrowths of staurolite and kyanite, while the lower part comprises larger garnet crystals in a muscovite matrix. The rock is somewhat altered and has veins of chlorite associated with the garnet crystals.

The views at higher magnification show a composite porphyroblast of epitaxially intergrown kyanite and staurolite in greater detail. It is cut to the right by a vein of chlorite, perhaps resulting from retrograde alteration along a crack. Although much of the matrix is coarse decussate muscovite, there is some retrograde chlorite and the porphyroblast is flanked on its upper margin by plagioclase.

The occurrence of parallel intergrowths of staurolite and kyanite is noted in most mineralogical texts and results from the similarity of parts of their structures. Nevertheless it is not very frequently observed.

Locality: Zion Hill, Ox Mountains, Co Sligo, Ireland.
Magnification: × 7, XPL; and × 20 PPL and XPL.

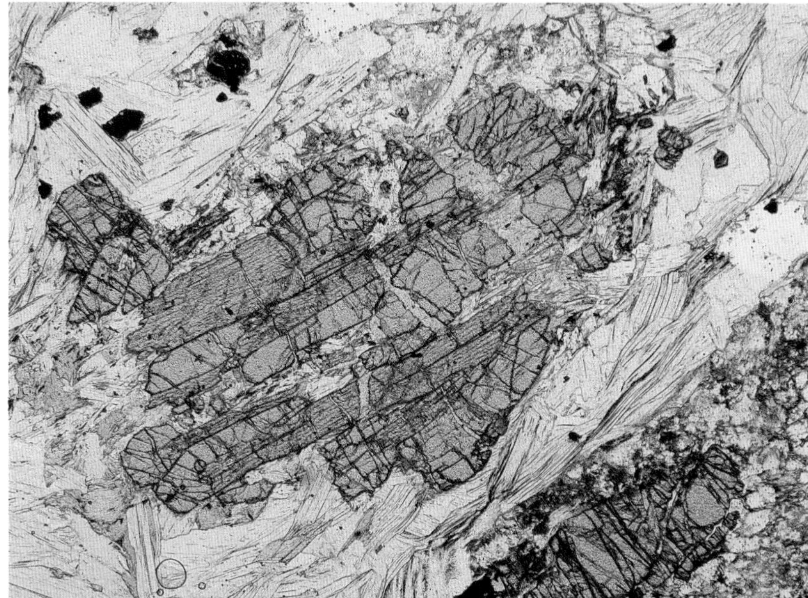

18

Kyanite biotite graphite schist

Amphibolite facies – kyanite zone

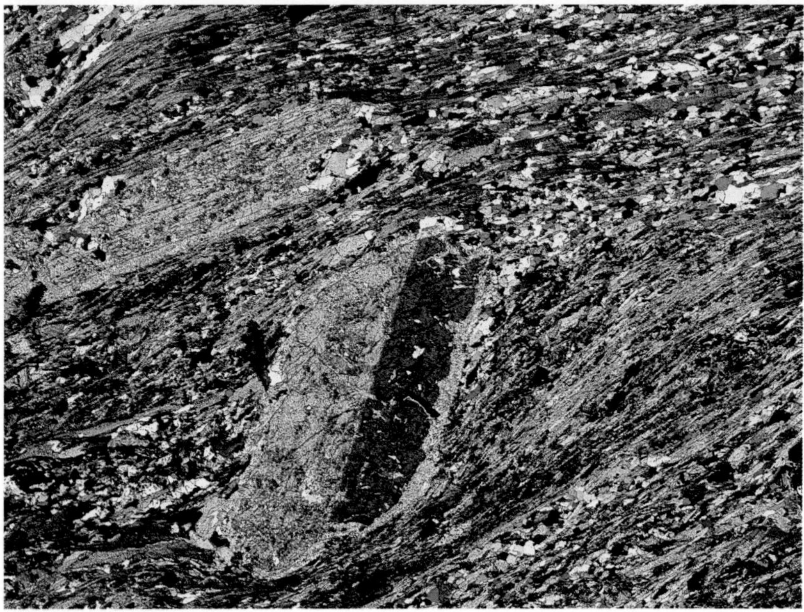

Two porphyroblasts of kyanite are shown in this view, one of them is simply twinned and both of them are surrounded by a retrograde *shimmer aggregate* of fine muscovite. The groundmass of the rock is mainly biotite, muscovite, graphite and quartz with rare crystals of tourmaline.

The dominant, diagonal mica foliation may itself have been produced by crenulation of an earlier fabric. Graphite in the lower left quadrant picks out numerous microfolds to which the dominant foliation is axial planar.

Locality: Chiwaukum schist, Stevens Pass, Northern Cascades, Washington, USA. Magnification: × *9, PPL and XPL.*

19

Sillimanite staurolite schist

Amphibolite facies – sillimanite zone
(Additional examples: **96, 100**)

The field of view shown in the lower magnification photographs reveals that the main minerals in this rock are staurolite, biotite, plagioclase and quartz. Several conspicuously zoned grains of tourmaline are also visible near the middle of the upper and right-hand edges (green cores, yellow rims), and there are some corroded remnants of an original garnet porphyroblast which are so heavily clouded as to appear nearly opaque in PPL. The garnet remnants are mantled by biotite which is intergrown with, and partially replaced by, fibrolitic sillimanite. This can be seen more clearly in the detailed high power view (some small air bubbles in the lower right part of the slide should not be confused with minerals). The replacement of garnet in this way leads ultimately to the development of sillimanite pseudomorphs after garnet (*see* **100**) as a result of a complex ionic reaction cycle.

Locality: Cur Hill, Connemara, Ireland. Magnification: × 20, PPL and XPL; and × 56, PPL.

High temperature metamorphism

In some parts of the world, for example the Appalachian belt in northeast USA, the Barrovian sillimanite zone is succeeded by progressively higher grade zones. The first evidence for this is the breakdown of muscovite + quartz → K-feldspar + sillimanite + fluid, and the appearance of migmatite leucosomes of broadly granitic compositions. The transition from the upper amphibolite facies to the granulite facies is marked by the coexistence of the four phases garnet + cordierite + K-feldspar + sillimanite. While in some areas extensive migmatites are developed under amphibolite facies conditions, elsewhere significant melting is restricted to the granulite facies. These contrasting styles of high temperature metamorphism are probably controlled by the availability of water. In a few parts of the world extremely high temperature metamorphism of pelites has taken place resulting in the formation of exotic mineral assemblages such as the coexistence of sapphirine + quartz.

20

Sillimanite K-feldspar biotite schist

Amphibolite facies – sillimanite K-feldspar zone
(Additional example: **84**)

This assemblage is rather typical of high grade schist in which the temperature has been sufficiently high for the muscovite to react with quartz to produce K-feldspar and an aluminium silicate mineral, in this case sillimanite. To make it possible to distinguish the K-feldspar easily from untwinned plagioclase or quartz the section has been stained with sodium cobaltinitrite solution after being etched with hydrofluoric acid vapour.

In the lower part of the PPL view the pale yellow, stained K-feldspar can be readily distinguished from yellow-brown biotite and quartz and plagioclase. In the upper part of the field of view fine needles of fibrolitic sillimanite are easily seen intergrown with quartz and biotite. A few muscovite crystals are probably of retrograde origin rather than relics from lower grade. The segregation of the fibrolitic sillimanite and K-feldspar into separate domains, although a common phenomenon at this grade, is not well understood.

In many terranes, melting precedes or accompanies muscovite breakdown. The absence of migmatitic features in this rock reflects rather low pressures of metamorphism.

Locality: Maumeen, Connemara, Ireland. Magnification: × 27, PPL and XPL.

21

Garnet cordierite plagioclase sillimanite gneiss

Amphibolite facies – sillimanite K-feldspar zone

The main minerals in this rock are listed in the name above and we can also include biotite and quartz.

The garnet and biotite are readily identified, while the fibrolitic form of sillimanite is developing mainly at the expense of biotite, notably in the centre of the field of view. Large crystals in the top left quadrant are pseudomorphs after cordierite which has been almost entirely replaced by fine sericite due to a retrograde reaction. Plagioclase shows incipient alteration in PPL (lower right quadrant), but still displays multiple twinning. Quartz is clear and unaltered.

The association of garnet, cordierite and sillimanite without K-feldspar is diagnostic of low to medium pressure metapelites in the uppermost part of the amphibolite facies. This rock is the schistose portion of a migmatite and is enriched in aluminous minerals due to melting (i.e. it is a restite). It is typical of areas where there has been extensive melting under upper amphibolite, as opposed to granulite, facies conditions.

Locality: Lake Nahasleam, Connemara, Ireland. Magnification: × 13, PPL and XPL.

22

Garnet cordierite K-feldspar gneiss

Granulite facies
(Additional example: **101**)

This rock consists mainly of microcline–perthite and quartz with concentrations of garnet, cordierite and a small amount of biotite and iron ore. The cordierite can be recognized by the very characteristic alteration to yellowish pinite seen in the upper part of the PPL view. The XPL view at higher magnification (a detail from the upper left quadrant of the lower power view) displays the isotropic veins and crack fillings replacing low birefringence cordierite which are very characteristic of this alteration.

The clear mineral speckled with inclusions in this rock is quartz, whereas the microperthite is free from tiny inclusions.

The assemblage of this rock is typical of lower granulite facies pelitic migmatites.

Locality: Kakola, Turku, Finland. Magnification: × 9, PPL and XPL; and × 25, XPL.

23

Migmatitic gneiss

Granulite facies

A granulite facies migmatitic rock consisting of restite melanosome of garnet, sillimanite, spinel, biotite, cordierite and an opaque oxide mineral alternating with coarse leucosomes containing K-feldspar plagioclase and quartz.

The higher magnification view is taken from the centre of the field of the lower power view and the garnet, prismatic sillimanite (with diagonal cleavage) and biotite can be readily identified. The cordierite tends to occur as rims around iron ore, as at the left hand edge of the higher powered view, near the upper left corner. Dark green spinel occurs in the lower right corner.

The leucosome is composed mainly of alkali feldspar and quartz but between this and the melanosome is a rim of plagioclase isolating the quartz from the spinel – the leucosome is likely to represent partially melted Si-rich material with the restite portion deficient in granitic components.

Locality: Kodaikanal, Southern India. Magnification: × 7, XPL; and × 22, PPL and XPL.

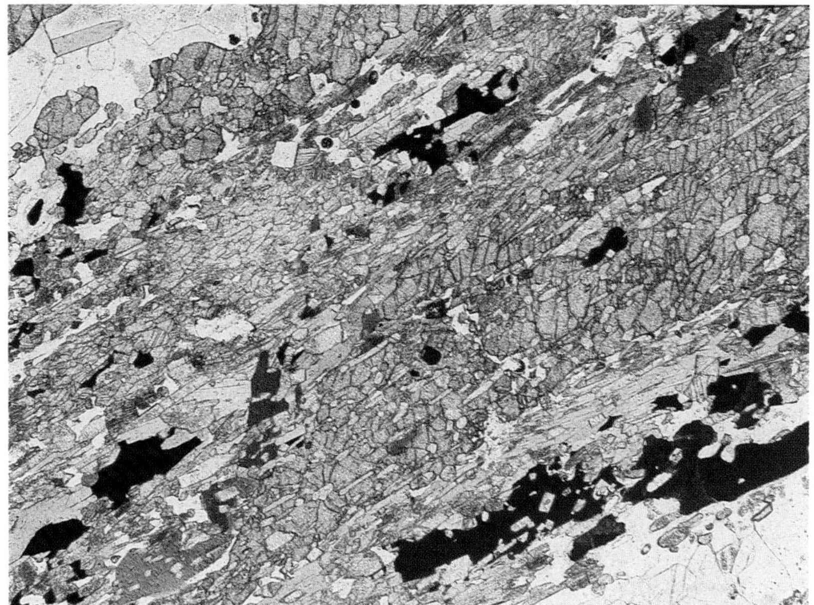

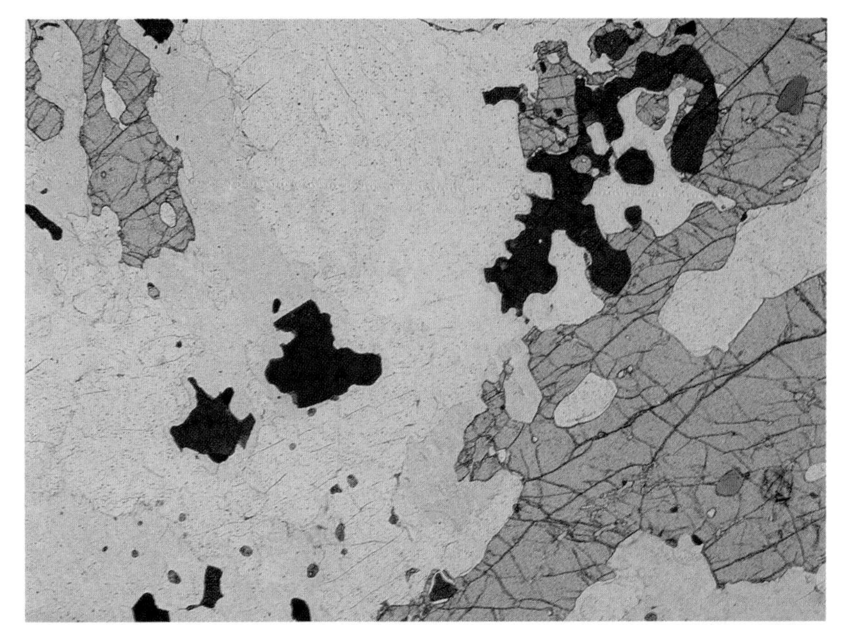

24

Garnet cordierite spinel quartz gneiss

Granulite facies

These photographs show a fairly coarse-grained rock in which garnet and a dark green (nearly opaque) spinel can be readily distinguished. The colourless minerals are microperthitic K-feldspar, plagioclase, cordierite and quartz. The cordierite is weakly *cloudy* in this rock due to myriads of small inclusions. A number of quartz crystals are in the extinction position and these have cracks filled with a micaceous mineral. Plagioclase displays multiple twinning, but no good examples of K-feldspar are visible in this view.

In the high magnification photograph a rim of cordierite can be seen armouring the spinel from contact with the quartz. This enlarged region is just to the right-of-centre of the lower power photograph in which the cordierite rim shows up in the XPL view as a white border to the spinel crystals.

Two biotite crystals can be seen within the garnet, these are the only hydrous minerals present, and have perhaps only been preserved under extremely high temperature conditions because they were armoured by the garnet.

Locality: 5 km west of Fort Dauphin, South Madagascar.
Magnification: × 16, PPL and XPL; and × 43, PPL.

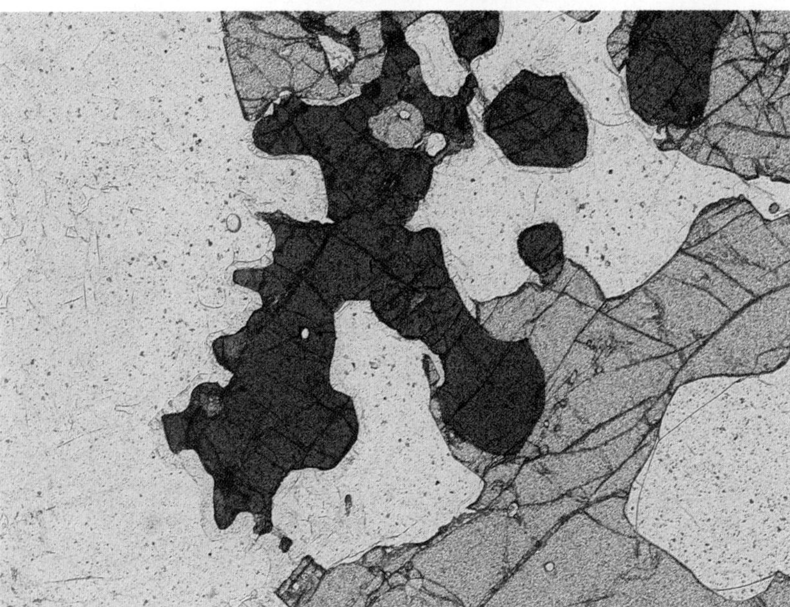

25

Sapphirine granulite

Granulite facies

The main minerals present in this rock are antiperthitic feldspar (not seen here), quartz and skeletal high relief sapphirine. Near the top of the field of view some orthopyroxene crystals showing first- to second-order interference colours, occur as rims around sapphirine crystals. The assemblage sapphirine + quartz is stable only at very high temperatures. At lower temperature the equivalent assemblage is orthopyroxene + sillimanite so that the orthopyroxene may have formed by a retrogressive reaction.

This assemblage is probably the highest temperature assemblage formed on a regional scale in metasediments. It requires temperatures in excess of 850 °C, and possibly around 1000 °C (*see also* **101**).

Locality: Enderby Land, Antarctica. Magnification: × 72, PPL and XPL.
Reference: Harley S L 1983 In Oliver R L, James P R, Jago J B (eds.) Antarctic Earth Sciences, Cambridge University Press, pp. 25–30

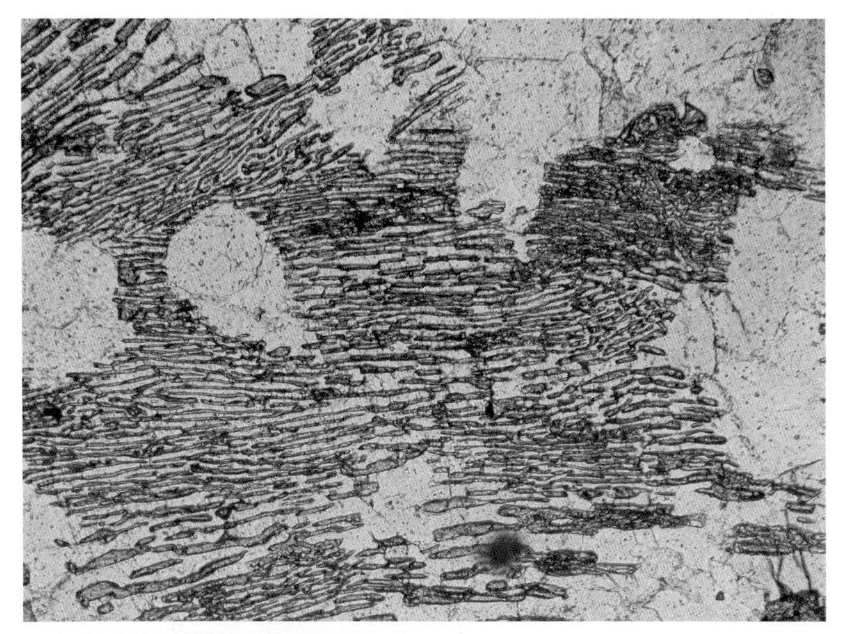

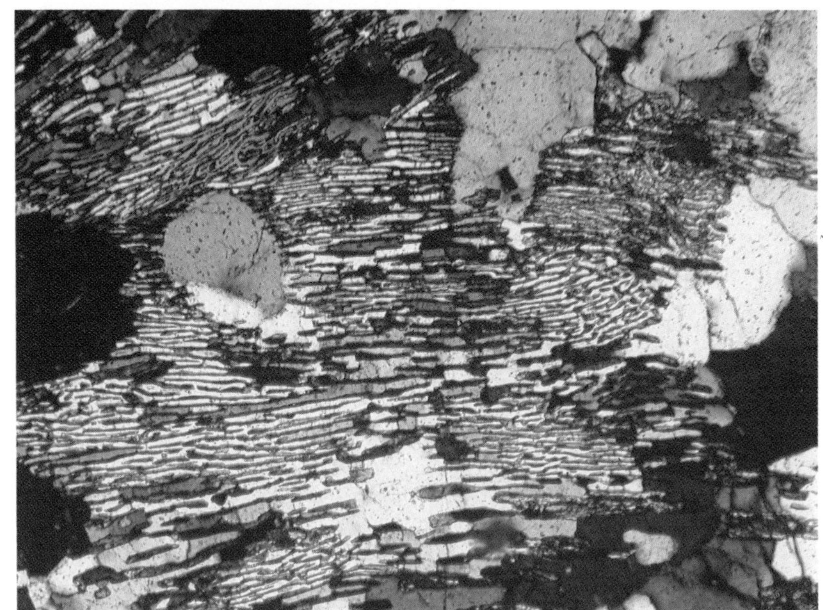

26

Sapphirine cordierite biotite gneiss

Granulite facies

This rock consists mainly of these three minerals. The skeletal sapphirine is intergrown with cordierite, which could easily be misidentified as plagioclase since it shows lamellar twinning and lacks some of its distinctive characteristics, such as pleochroic haloes or alteration to pinite.

A number of moderate relief grains are of apatite.

Locality: Europe claim, Beitbridge, Zimbabwe. Magnification: × 20, PPL.
Reference: Droop GTR 1989 Journal of Metamorphic Geology 7: 383–403

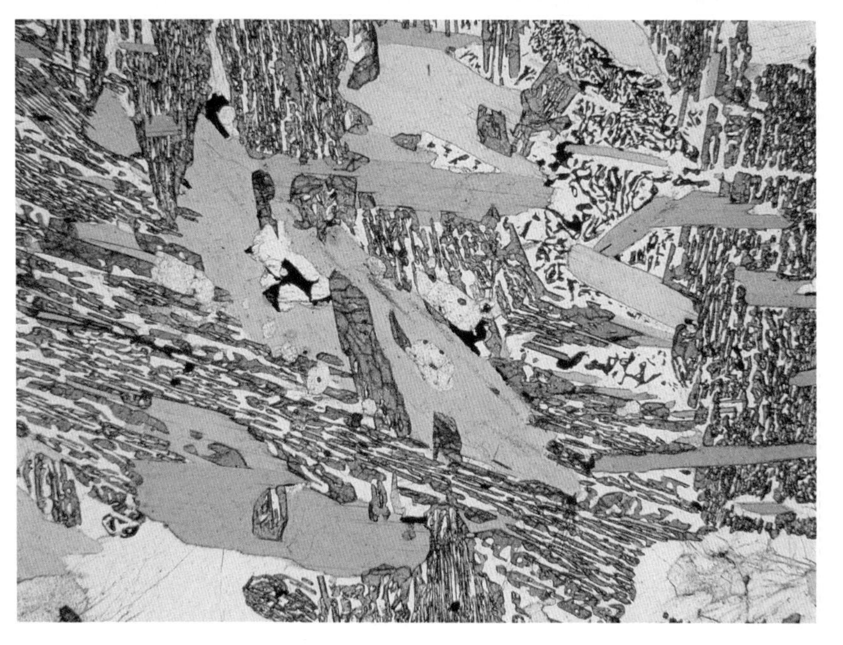

Effect of pressure variation on pelite assemblages

Low pressure metamorphism

In the lower pressure parts of the greenschist and amphibolite facies, pelitic schists and hornfelses develop andalusite rather than kyanite, while garnet becomes rare or absent and cordierite appears at progressively lower temperatures with a drop in pressure. At the lowest pressures, biotite hornfelses are succeeded by *spotted hornfelses* containing poikiloblastic cordierite (*see* **1**) while andalusite appears subsequently at higher grade. A number of examples of very high temperature metamorphism of pelite at near-surface pressures have been described from the vicinity of basaltic complexes. Here, wholesale melting of pelite, especially in xenoliths, may take place.

27

Andalusite (chiastolite) hornfels

Hornblende hornfels facies
(Additional examples: **1, 106**)

Two porphyroblasts of andalusite (variety chiastolite) are shown in this view and each has a rim of *shimmer aggregate* (probably muscovite). The andalusites are characterized by a pattern of graphite inclusions which has been compared to a Maltese cross. In some examples we find that although the original crystals of andalusite may have been completely replaced by fine micas, the pattern of inclusions remains. Generally the centres of the crystals are full of inclusions; in some cases, however, the centre of the cross may be free of them.

Despite the intense hornfelsing, the original slaty fabric and grain size is still apparent in the groundmass, which contains quartz, chlorite, biotite, muscovite and graphite.

Locality: Evans Lake aureole, Okanogon Co, Washington, USA. Magnification: × 14, PPL and XPL.

28

Cordierite andalusite hornfels (spotted slate)

Hornblende hornfels facies

This rock displays the characteristic appearance of a spotted slate (although this view has a greater density of spots than in many other examples). The spots are composed of two minerals *viz* andalusite and cordierite and in the PPL view they can be distinguished fairly easily since the andalusite crystals have higher relief than the cordierite; here they are also relatively free from inclusions. Three andalusite crystals are in the centre of the field while the other spots are of cordierite. Some of the cordierite crystals show sector twinning—the crystal above the centre of the field shows two sectors which are almost black and two are dark grey.

The presence of andalusite and absence of chlorite shows that this rock represents a higher metamorphic grade than sample **1**, from the same aureole, but what remains of the muscovite-rich matrix is nonetheless still very fine-grained.

Locality: Skiddaw aureole, England. Magnification: × 20, PPL and XPL.

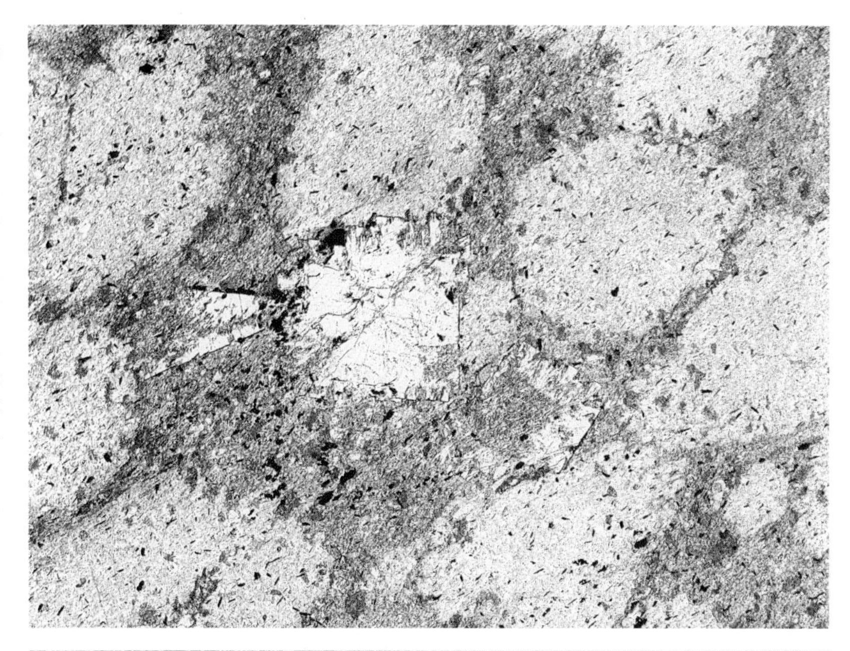

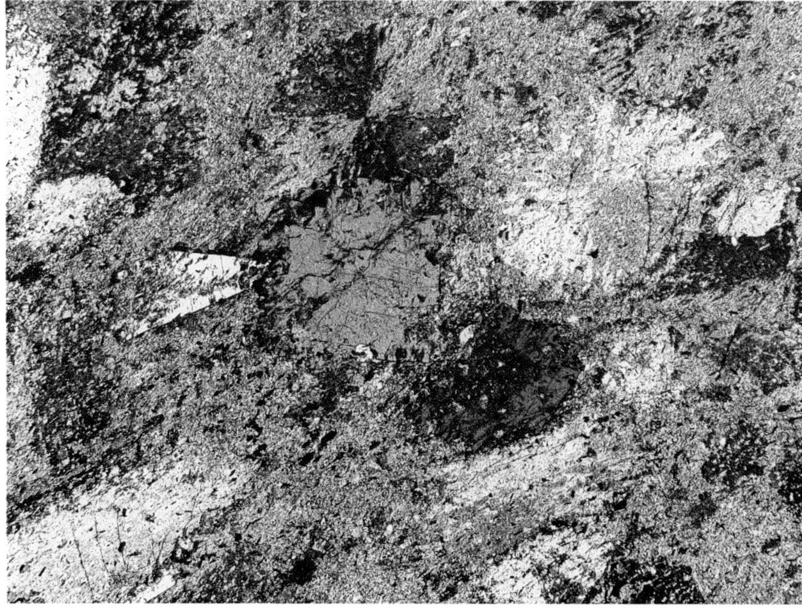

29

Andalusite biotite schist

Hornblende hornfels facies

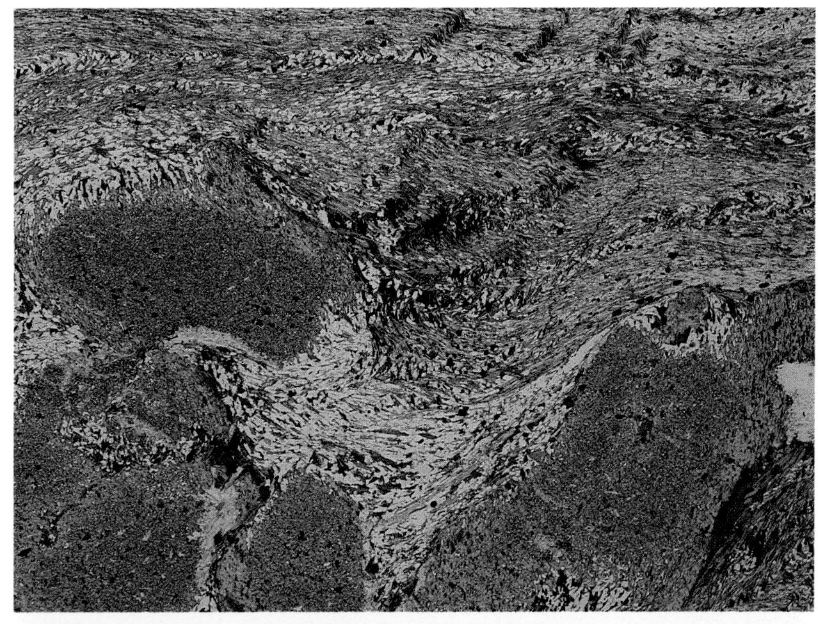

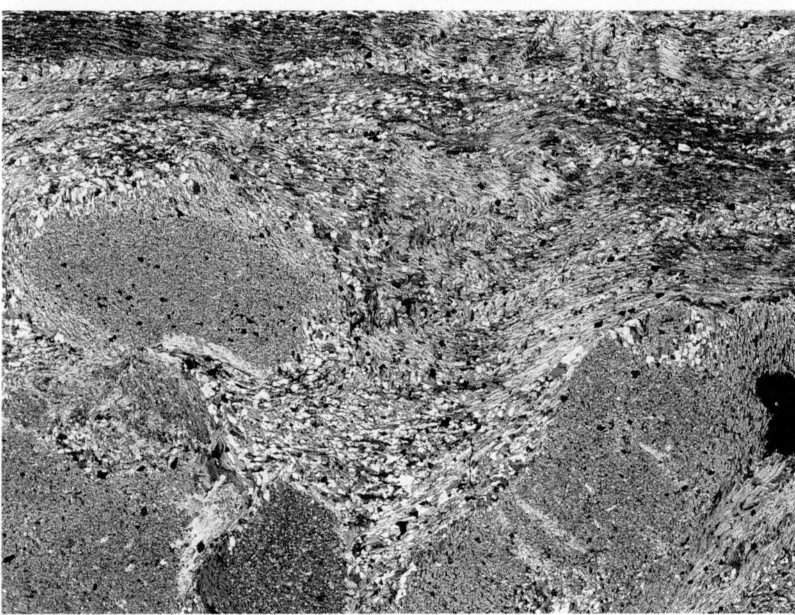

Large poikiloblasts of andalusite are set in a groundmass composed mainly of greenish-brown biotite, muscovite and quartz in this low pressure regionally metamorphosed rock.

The inclusions within the andalusites are distinctly finer-grained than the rock matrix and in some cases (e.g. in the lower right corner) define a N–S fabric that is at a high angle to the dominant E–W schistosity. In detail, it can be seen in the central and upper parts of the field of view that the E–W fabric is a crenulation schistosity produced by refolding of the early N–S foliation. This process has been accompanied by segregation into phyllosilicate-rich and quartz-rich bands. A final phase of deformation has produced kinks in the E–W schistosity near the upper edge of the field of view.

Locality: Black Water River, 1.5 km southwest of Bridgend, Grampian Region, Scotland. Magnification: × 8, PPL and XPL.

30

Andalusite staurolite schist

Hornblende hornfels facies

This rock consists of large poikiloblasts of andalusite with staurolite in a groundmass of biotite, with finer-grained muscovite and quartz. There is no clear evidence of any feldspar.

The poikiloblasts of staurolite are much smaller than those of andalusite and appear very dark in the PPL photograph. Just to the left-of-centre of the field of view is a poikiloblast that has been almost completely replaced by pale yellow pinite and may have been of cordierite originally.

Although described as a schist because the general mass of the rock has a schistose structure, the schistosity is not well defined in thin section.

Locality: Whitehills, near Banff, Scotland. Magnification: × 8, PPL and XPL.

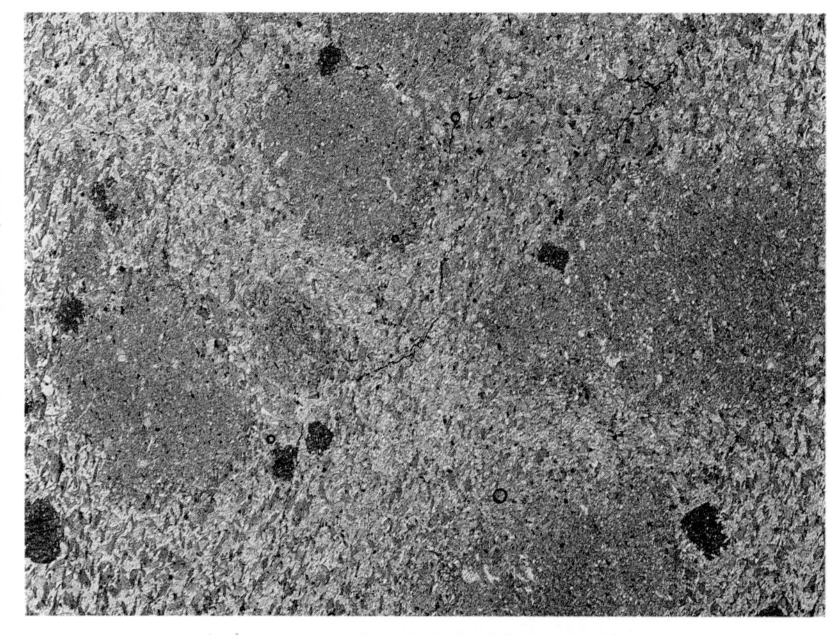

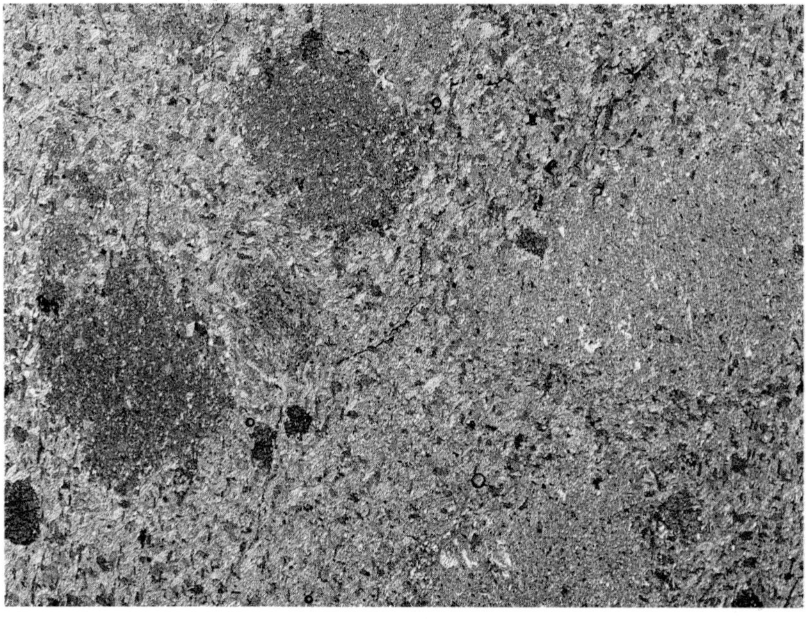

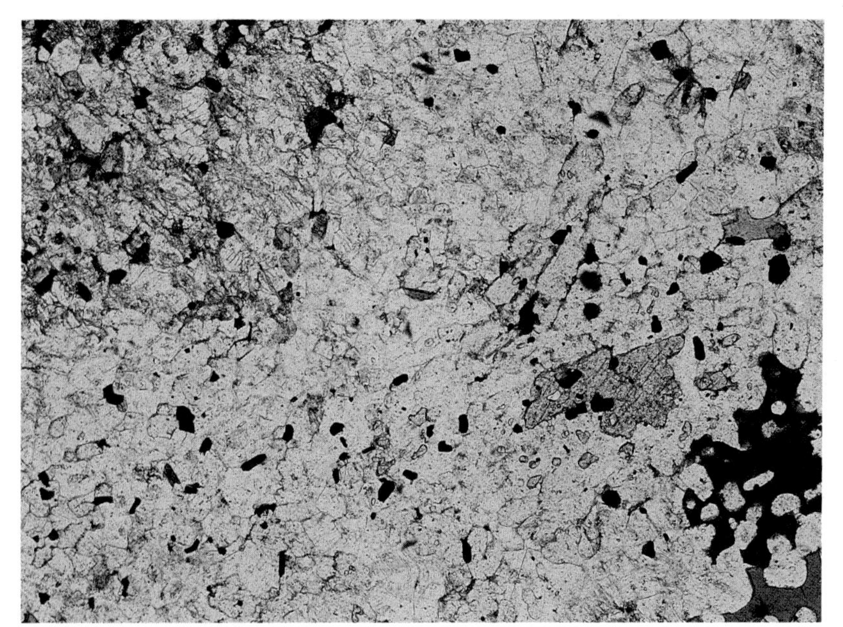

31

Andalusite cordierite K-feldspar hornfels

Pyroxene hornfels facies

This is a fine-grained rock consisting mainly of cordierite, andalusite, alkali feldspar and quartz. Cordierite is very difficult to distinguish, but sometimes, as around the central skeletal andalusite crystal in the high powered view, it shows lamellar twinning. Yellow pleochroic haloes are also seen in some cordierites. The alkali feldspar has a microperthitic texture and this aids in identifying it by giving a distinctive pattern of fine parallel lines or simply a patchy appearance in CPL.

Biotite and magnetite are present in small amounts. Minor muscovite is probably of retrograde origin.

The association of andalusite and K-feldspar results from breakdown of muscovite with quartz at very low pressures where andalusite, rather than sillimanite, is stable.

Locality: aureole of Ben Nevis granite, Scotland. Magnification: × 26, PPL and XPL; and × 52, XPL.

32

Cordierite plagioclase corundum spinel hornfels

Sanidinite facies

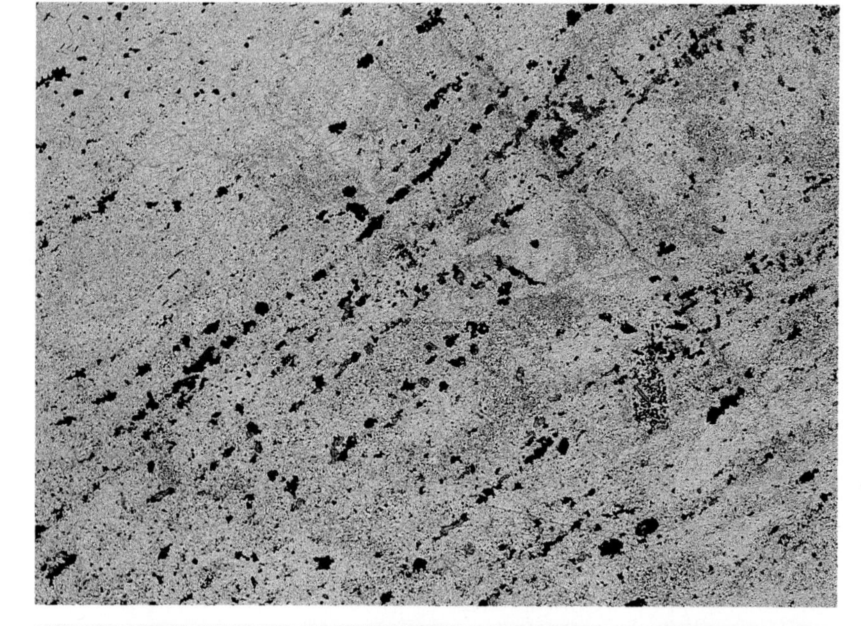

This is a very fine-grained metasediment in which the sparse small high relief grains with first-order yellow birefringence colours are corundum and the crystals which appear opaque in the low power view are in fact dark green spinels. The original layering is visible in the PPL photograph and in the XPL view large poikiloblastic cordierites are seen to occur throughout. Between the patches of cordierite in the upper left of the low powered view there are veins of what is probably alkali feldspar. Some plagioclase feldspar is intergrown with the cordierite but it is difficult to estimate the relative proportions of these two minerals.

This rock comes from an ultrametamorphosed xenolith. The high content of Al-rich minerals shows that it was formerly of pelitic composition, but high temperatures have destroyed all hydrous phases and it has been depleted in silica and alkalis by melting.

Locality: Invergeldie Burn, Glen Lednock, Comrie, Scotland. Magnification: × 8, PPL and XPL; and × 34, XPL.

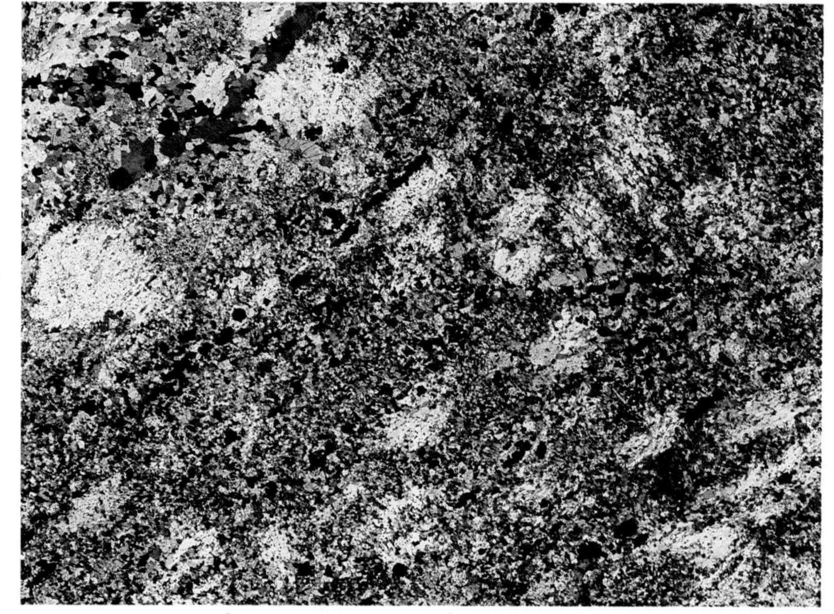

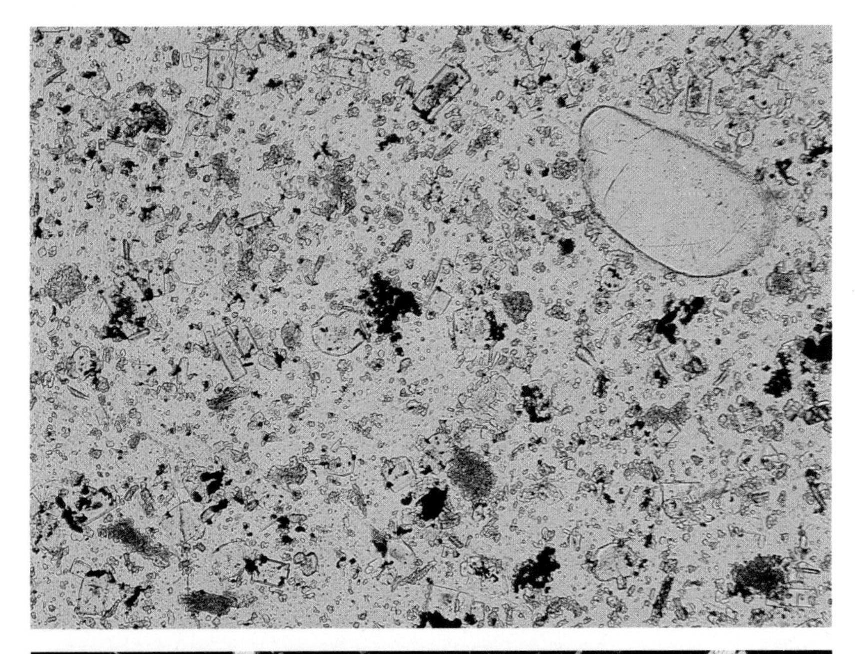

33

Buchite

Sanidinite facies

This name originally denoted a glassy rock formed by fusion of sandstone by an igneous rock but was later extended to include fused aluminous clays. This specimen contains rectangular, newly-formed crystals of cordierite and smaller laths of plagioclase. One partially resorbed relic crystal of quartz remains at the top right of the field of view. Other minerals present are orthopyroxene and possibly needles of mullite.

Locality: Cushendall, Co Antrim, Northern Ireland. Magnification: × 52, PPL and XPL.

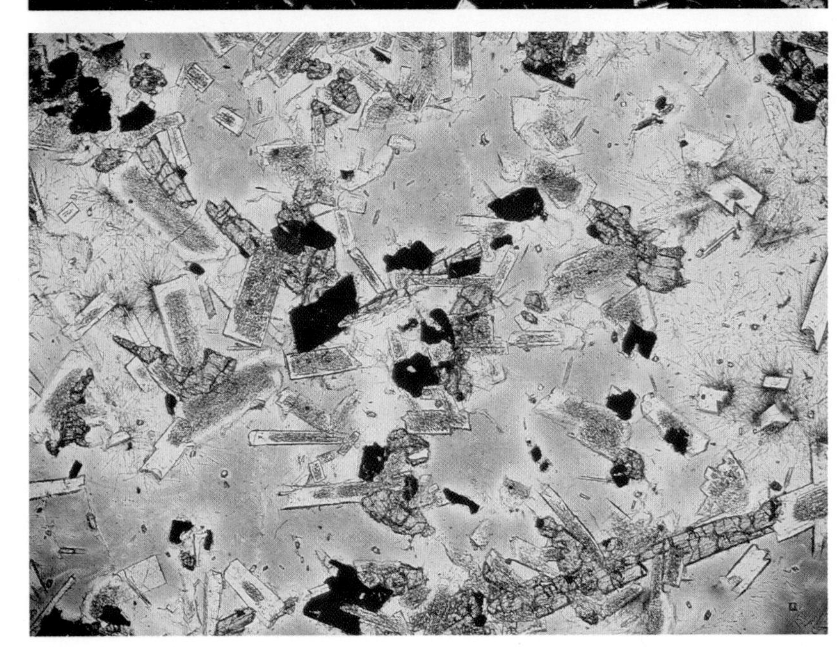

34

Buchite

Sanidinite facies

This is another example of a glassy rock formed by melting of sediment at a lava contact. The minerals present in this view all crystallized from the melt and are dominated by plagioclase, cordierite, orthopyroxene and mullite. Both plagioclase feldspar and cordierite have moderate relief and are crowded with inclusions making them difficult to distinguish. The high relief crystals are of orthopyroxene and the minute needles in the colourless areas of glass are mullite. The opaque phases are ilmenite and magnetite.

Locality: near Cushendall, Co Antrim, Northern Ireland. Magnification: × 34, PPL.

High pressure metamorphism

The effects of high pressures on pelite assemblages have been less well known until recent years, because the pelites of most high pressure metamorphic belts come from less mature sedimentary environments than most Barrovian pelites. Recent work, notably in the European Alps, has however identified a number of distinct high pressure minerals and mineral assemblages. These include the occurrence of carpholite (**35**) and the coexistence of talc with phengite muscovite or, at higher temperature, kyanite (**36**).

Further details of the metamorphic assemblage and reactions of pelitic schists are given in Yardley (1989, Chapter 3).

35

Carpholite chloritoid schist

Blueschist facies

This is a rather fine-grained metamorphosed argillaceous rock within a Triassic limestone series. The minerals it contains are Mg-rich carpholite (about 70% of the Mg end-member) chloritoid, phengitic mica, calcite and small amounts of chlorite and quartz.

The distinctive mineral indicative of unusually low temperature and high pressure metamorphism is the Mg–Fe carpholite. This occurs as bundles of near-parallel prisms with only moderate relief (similar to that of muscovite). Where the prisms are cut parallel to their length they display low first-order grey birefringence colours (as at the centre of the field of view). Oblique and basal sections tend to be lozenge shaped and show bright first-order birefringence colours ranging up to second-order blue. The smaller radiating aggregates of crystals of much higher RI than the carpholite are of chloritoid. The rock matrix is predominantly of phengite with lesser amounts of calcite and quartz.

Locality: Western Vanoise, Dent de la Portetta, Western Alps. Magnification: × *27, PPL and XPL.*
Reference: Goffe B, Velde B 1984 Earth and Planetary Science Letters **68**: 351–60.
Goffe B 1980 Bulletin de Minéralogie **13**: 297–302.

36

Talc kyanite schist (whiteschist)

Eclogite facies

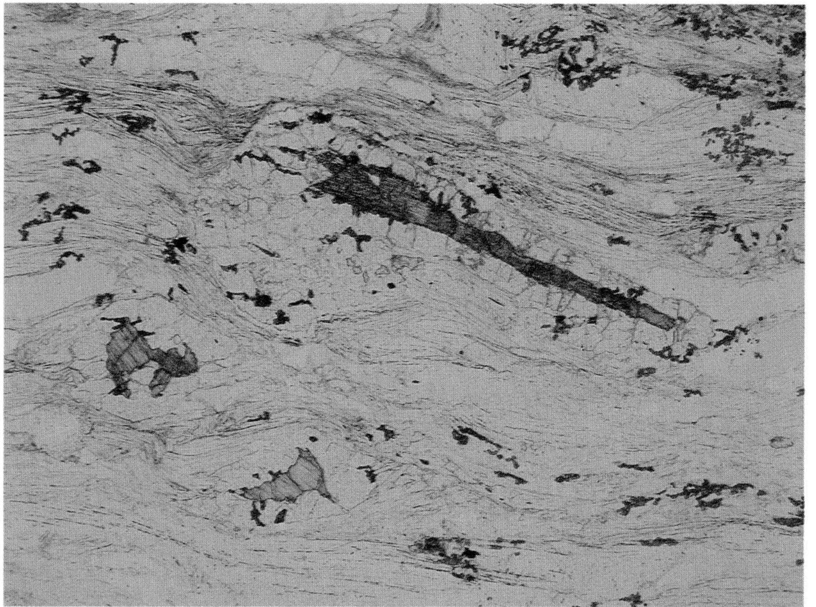

This view shows an elongated crystal of kyanite and at the bottom left of the field of view are two other kyanite crystals. The low birefringence mineral surrounding each of the kyanite crystals is cordierite produced by retrograde depressurization.

The micaceous mineral occupying a large part of the field of view, and showing bright second-order interference colours, is talc; this cannot easily be distinguished from muscovite in thin section. The rest of the field of view is made up mainly of quartz.

The assemblage of talc–kyanite is an indicator of high pressure and in the presence of excess quartz it reverts to cordierite at higher temperatures and lower pressure.

The name *whiteschist* was adopted by W Schreyer to describe the facies of rocks formed under conditions where talc + kyanite are stable.

Locality: Sar e Sang, Afghanistan. Magnification: × *20, PPL and XPL.*
Reference: Kulke H, Schreyer W 1973 Earth and Planetary Science Letters **18**: 824–8

37

Pyrope kyanite talc phengite schist with coesite

Eclogite facies

This is a metasedimentary rock from a high grade eclogite facies region. It is characterized by having pale garnets varying in size from 0.2 to 25 cm in diameter. These photographs are of one of the garnets surrounded by talc, kyanite, phengite and quartz. Inclusions in the garnet are mostly kyanite and quartz, but the large low relief quartz inclusions contain high relief remnants of its denser polymorph coesite. The quartz in these inclusions also displays a curious texture that is characteristic of pseudomorphs after coesite. Radial cracks in the garnet around these inclusions have been caused by the large volume increase on inversion of the coesite to quartz and this may have happened at relatively low temperature during uplift.

Locality: Dora Maira massif, Western Alps. Magnification: × 25, PPL and XPL.
Reference: Chopin C 1984 Contributions to Mineralogy and Petrology **86**: 107–18

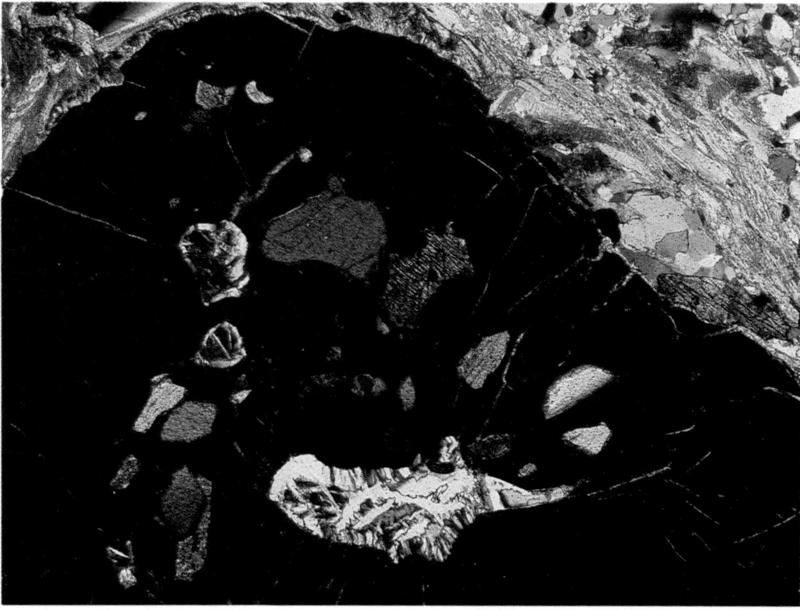

38

K-feldspar kyanite granulite

Granulite facies

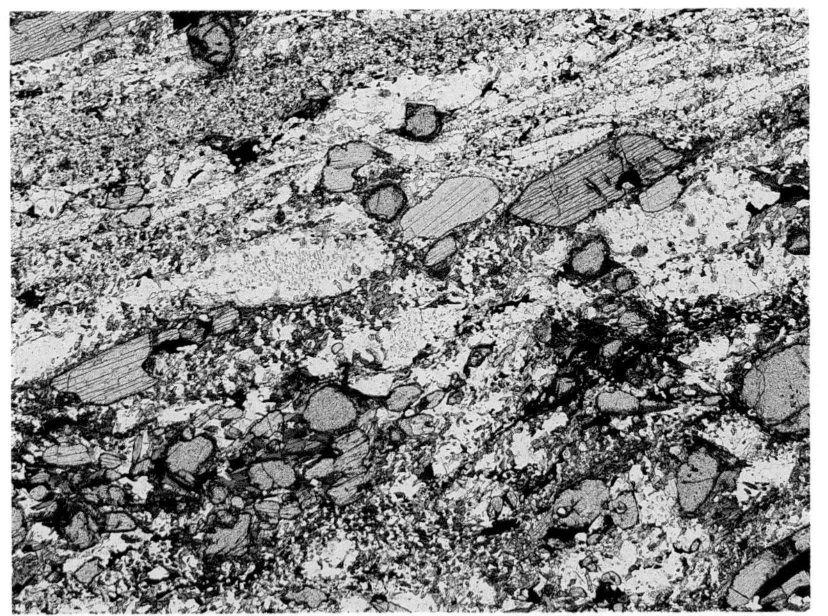

This rock is a mylonitized and recrystallized granulite. The thin section is slightly thick so that kyanite crystals have somewhat high interference colours. In addition to kyanite, the main minerals present in this rock are garnet, biotite, perthitic K-feldspar and quartz. To the left-of-centre of the field we can see a K-feldspar porphyroclast containing exsolved blebs of plagioclase in its core while its margins have recrystallized to polygonal, non-perthitic K-feldspar. The quartz is largely fine-grained.

The presence of K-feldspar together with kyanite is an indication of the breakdown of muscovite with quartz in the kyanite stability field (cf. **20**, **31**). The fact that the deformation rounded and corroded kyanite and garnet and caused perthite to break down as it recrystallized demonstrates that the mylonitization is a later, lower temperature event, after the peak of metamorphism.

Locality: Slishwood, Co Sligo, Ireland. Magnification: × 12, PPL and XPL.

Metamorphism of tuffs, greywackes and cherts

The lithologies illustrated in this chapter are largely absent from the metamorphic rocks of the Caledonian–Appalachian belt in which many early classic studies were carried out, but prove to be valuable metamorphic indicators in very low grade and high pressure metamorphic belts. Indeed the zeolite facies was first erected by D S Coombs (1954) on the basis of the assemblages in metagreywackes from New Zealand.

Volcanogenic greywackes develop metamorphic assemblages even at very low temperatures because they contain highly reactive fragments of glass and igneous minerals whilst retaining, at least initially, the porosity of a sandstone. Hence the igneous materials break down very soon after burial to produce low temperature zeolite minerals. At higher temperatures, assemblages are probably very similar to those of other metamorphosed igneous rocks of comparable composition; it is their unique reactivity that makes greywackes valuable low grade indicators.

Cherts (**44–46**) and ironstones (**47–48**) display an even greater diversity of composition and assemblages than greywackes. While all cherts are, by definition, rich in silica, some have high levels of Fe (**44**, **45**) while others are Mn-rich (**46**) and develop minerals close to the Mn end-members of Fe–Mn solid solutions.

39

Laumontite metagreywacke

Zeolite facies

This rock has been subjected to rather low grade metamorphism and most of its characteristic sedimentary features are still visible. It must have originally consisted of a poorly sorted collection of angular volcanogenic fragments of feldspar and quartz together with ferromagnesian minerals which have since been replaced by secondary material rich in chlorite but stained by ferric iron. These fragments are embedded in a groundmass too fine-grained for optical determination. In PPL the clearest mineral fragments are of quartz, whereas the feldspar has been partly replaced by laumontite. In the lower left quadrant one nearly rectangular fragment of clouded feldspar is partly replaced by clear polycrystalline laumontite at its lower left corner.

Laumontite is distinguished from other zeolites by being biaxial negative and having a low optic axial angle.

Locality: Jurassic sediments, near Ship Cove, Hokonui Hills, New Zealand. Magnification: × 72, PPL and XPL.
Reference: Boles J R, Coombs D S 1975 Geological Society of America Bulletin **86**: 163–73

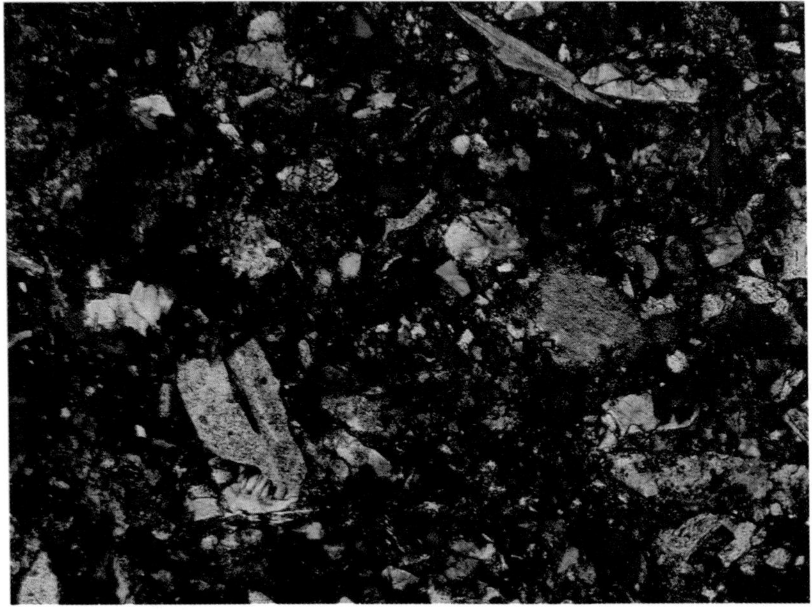

40

Heulandite meta-tuff

Zeolite facies

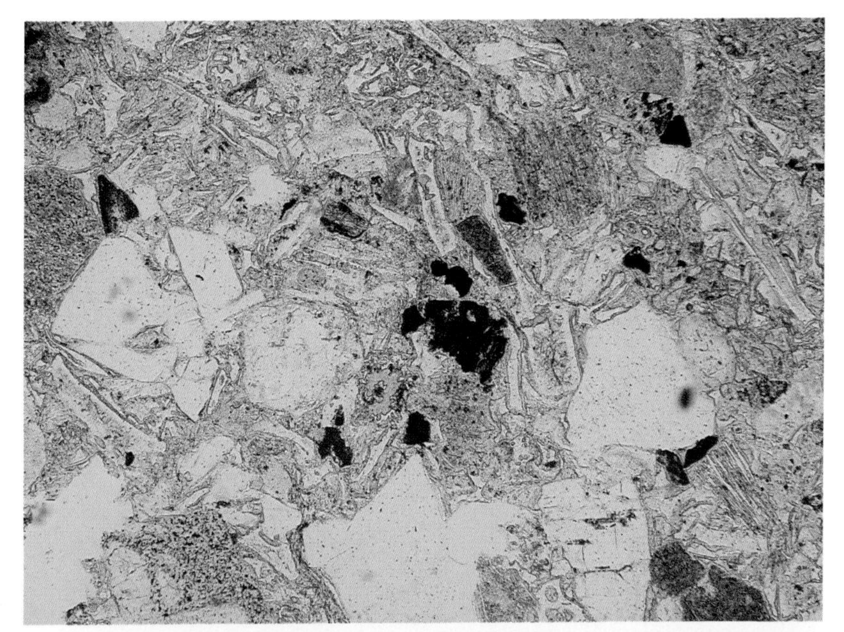

In addition to crystalline particles, this tuff originally contained abundant glass shards. Angular fragments of both individual feldspar crystals and fine-grained volcanic rock (seen at the left hand edge) are almost unaltered. However the fine-grained groundmass is partly replaced by green chlorite and individual elongate glassy shards are outlined by rims of chlorite. The interior portions of the shards are replaced by fine-grained aggregates of the zeolite heulandite. Secondary calcite is also present.

The absence of deformation is typical of rocks subjected to burial metamorphism.

Locality: North Range, South Island, New Zealand. Magnification: × *53, PPL and XPL.*

41

Jadeite glaucophane metagreywacke

Blueschist facies

This rock, originally a greywacke, was metamorphosed to produce the jadeite–glaucophane assemblage and is now weakly foliated. One glaucophane crystal, readily apparent from its blue colour, occurs above the centre of the field of view, but most of the other high relief material is jadeite, forming 20–30% of the rock. It occurs like glaucophane as bundles of radiating crystals, and has low birefringence. Much of the rest of the rock is quartz, including relic detrital grains, and there is minor phengite.

Locality: Panoche Pass, California, USA. Magnification: × 20, PPL and XPL.
Reference: Ernst W G 1965 Geological Society of America Bulletin **76**: 879–914

42
Pumpellyite actinolite schist
Prehnite pumpellyite facies

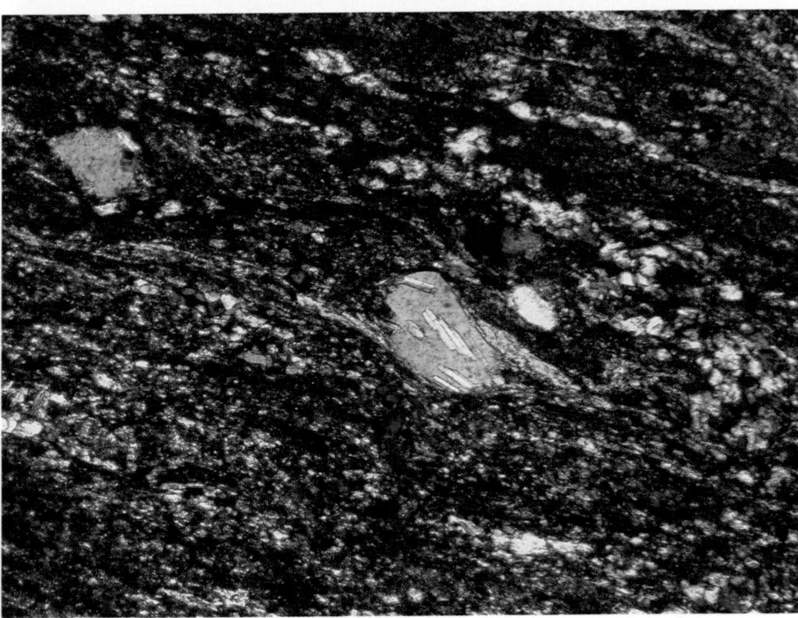

This is an extremely fine-grained rock and mineral identification is not easy. The bulk of the rock is composed of fine-grained quartz with chlorite and minor epidote. Sporadic bands parallel to the schistosity are of coarser pale actinolite with distinctive orange, red and blue birefringence colours. Relatively large low relief areas are of albite and one albite porphyroclast near the centre of the field of view contains several elongate pale grains of pumpellyite.

Locality: south of lookout, Baronet's Bluff, west of Queenstown, New Zealand. Magnification: × 72, PPL and XPL.

43

Stilpnomelane metagreywacke

Prehnite pumpellyite facies

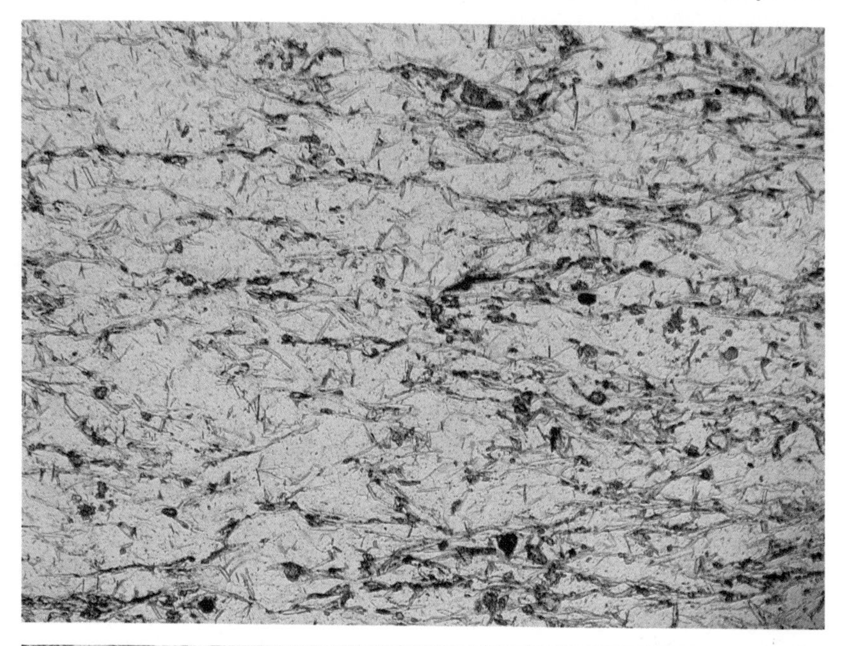

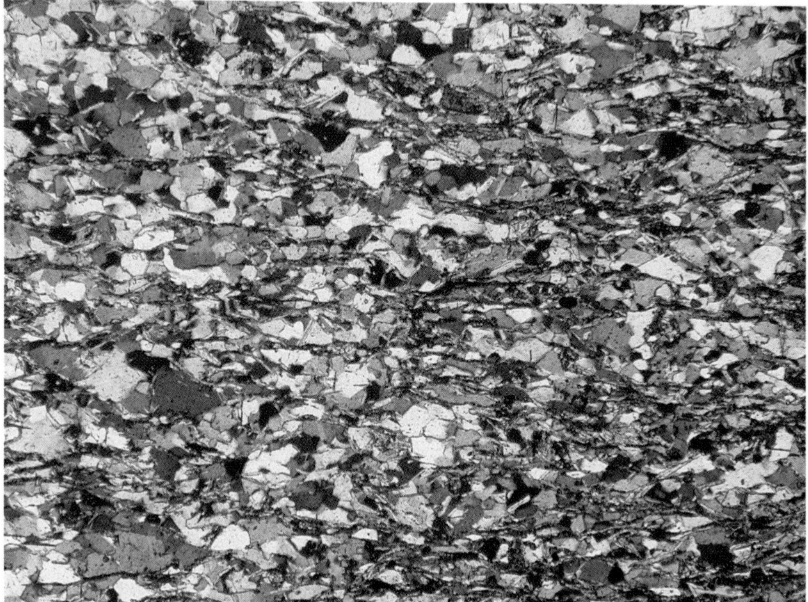

This is a very fine-grained rock containing muscovite, stilpnomelane, an epidote mineral, quartz and albite, with a small amount of tourmaline. The major minerals can be distinguished in the higher magnification view. Stilpnomelane forms characteristic thin bladed grains that are strongly coloured greenish-brown. Pale green phengitic muscovite has a similar habit and is often associated with small high relief granules of epidote. Two epidote grains can be distinguished by their relief and birefringence along the bottom edge of the low powered views. Rare high relief blue-green crystals are tourmaline (visible at high power only, with difficulty).

Locality: Lake Hawea, New Zealand. Magnification: × 12, PPL and XPL; and × 72, PPL.

44

Stilpnomelane schist

Greenschist facies
(Additional example: **98**)

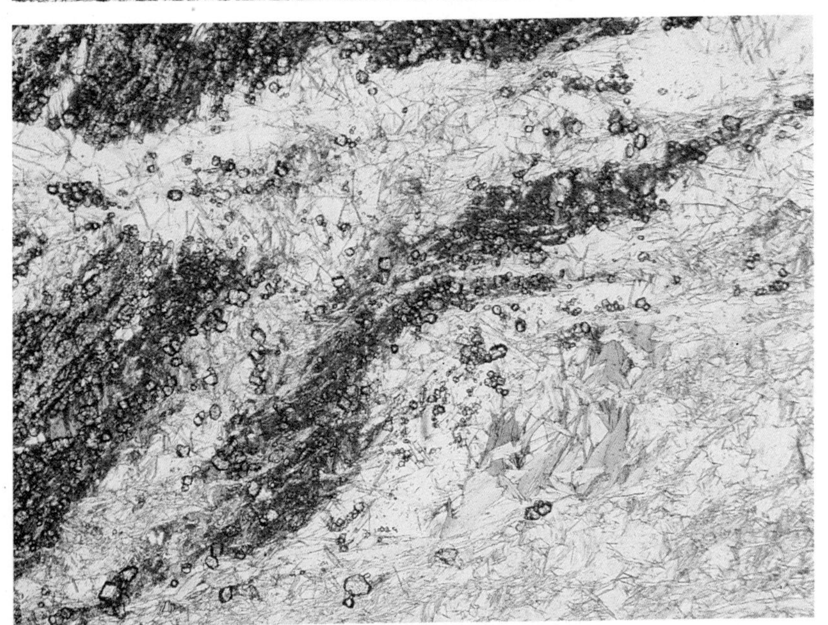

This is a rock rich in green ferrostilpnomelane, with chlorite, epidote, muscovite, quartz and garnet. The low power view shows a fold outlined by a band of opaque minerals and chlorite, enclosing a quartz-rich part of the rock. The high magnification views are taken from the top left of the low power field of view and show the stilpnomelane needles in quartz. Fine muscovite is also present. A high concentration of small high relief grains is also apparent at this magnification and these are mostly spessartine garnet. However not all are isotropic and some are of epidote.

Green, ferrous iron bearing ferrostilpnomelane is the stable form of stilpnomelane under metamorphic conditions, but it rapidly weathers to brown ferristilpnomelane once close to the surface.

Locality: Queenstown, New Zealand. Magnification: × 10, XPL; and × 53, PPL and XPL.

45

Riebeckite aegirine-augite metachert

Blueschist facies

This quartz-rich rock contains several distinctive metamorphic minerals set in a matrix of deformed and syntectonically recrystallized quartz, with phengitic muscovite.

Sodic amphibole occurs as large, zoned grains with low birefringence. Cores are of pale magnesioriebeckite, whereas the Fe^{2+}-enriched riebeckite rims appear deep blue in suitably oriented grains. Aegirine-augite is finer grained, pale green and of high relief. It is best seen near the centre of the XPL field of view where its bright birefringence colours, closely comparable to epidote, are distinctive. Spessartine garnet occurs as small euhedral grains throughout the rock, the fine grain size being typical of manganiferous garnets in metacherts. A single grain of moderate relief apatite occurs in the centre of the upper part of the field of view.

Locality: Bizan, Tokushima Prefecture, Japan. Magnification: × 50, PPL and XPL.
Reference: Miyashiro A, Iwasaki M 1957 Journal of the Geological Society of Japan **63**: 698–703

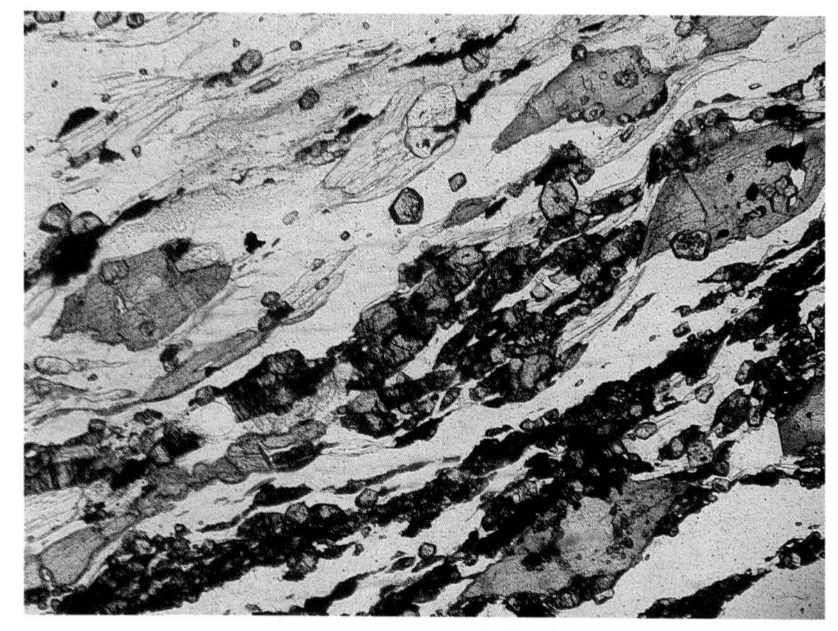

46

Piemontite metachert

Blueschist facies

Mn-rich but Fe-poor cherts are widespread in circum-Pacific metamorphic belts as well as in the Alps, Cyclades and elsewhere, and may provide useful indicators of metamorphic grade in otherwise monotonous sequences.

In the example shown here the Mn-mineral is piemontite, the manganese epidote. Pink colourations are common amongst manganese minerals, but in piemontite the pleochroism is particularly marked in shades of red, pink and yellow. The matrix is of quartz and phengite.

Locality: Karystos, South Evia, Greece. Magnification: × 65, PPL.

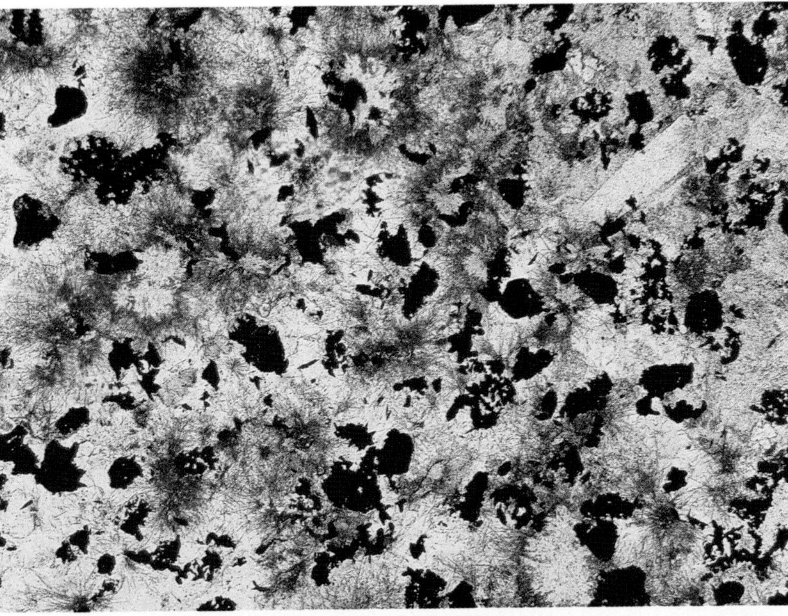

47

Minnesotaite grunerite meta-ironstone

Greenschist facies

This rock was originally a ferruginous chert which underwent low grade regional metamorphism followed by contact metamorphism. Over most of the field of view the rock is composed of magnetite, quartz and yellow radial aggregates of fibrous minnesotaite, the Fe end-member of talc. Note the more conspicuous relief of the minnesotaite fibres where they extend out into quartz. In the upper right quadrant are a few coarse grains of colourless, brightly birefringent grunerite. Minor carbonate occurs near grunerite and to the left of the field of view. It is best seen in crossed polars and is probably siderite.

Locality: Erie Mine, Mesabi Range, Minnesota, USA. Magnification: × 25, PPL and XPL.
Reference: French BM 1968 Minnesota Geological Survey Bulletin **45**

48

Grunerite magnetite quartzite

Amphibolite facies

This is a good example of a medium grade metamorphosed ironstone. It shows superimposed on the original bedding a pronounced alignment of the iron minerals magnetite and grunerite. The enlarged view has been taken from a small area on the lower edge of the left side of the field of view and the multiple twinning, which is a characteristic of grunerite, can be readily seen. The main constituent of the rock is quartz so that it was probably originally a chert. It is from a Precambrian banded iron formation.

Locality: Dwala Ranch, Gwanda District, southern Zimbabwe. Magnification: × 12, PPL; and × 34, XPL.

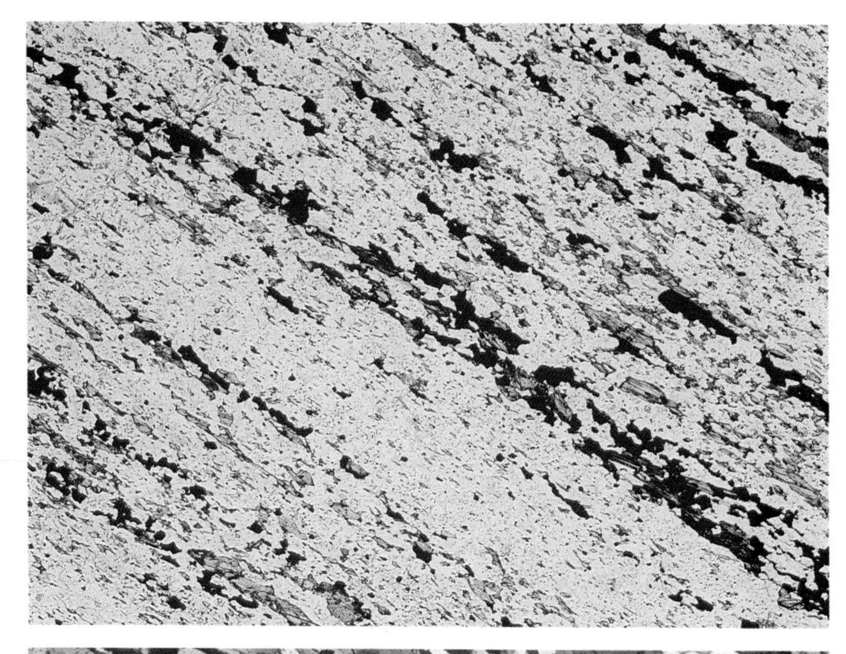

Metamorphism of marbles and calc-silicate rocks

Both these rock types are distinguished by the presence of Ca-rich minerals including Ca–Mg silicates whose compositions are rich in Mg relative to Fe. Marbles contain abundant carbonate (usually calcite and less commonly dolomite; rarely other carbonate minerals may be important), whereas in calc-silicates the carbonate is subordinate and may be absent. The distinction between the two rock types is however a gradational one. Many calc-silicates result from metamorphism of impure calcareous sediments such as marls, however others are probably of metasomatic origin, formed by interactions between original thin limestone layers and adjacent pelite.

Calcite limestones with quartz sand as the principal impurity react little during metamorphism unless there are extreme conditions of pressure (calcite is replaced by aragonite) or temperature (wollastonite may form if the pressure is also low). However extensive textural changes take place during metamorphism of marbles even where no mineralogical reaction takes place. Many limestones are dolomitic however, and they are much more reactive during metamorphism in the presence of silica. At low grades, talc appears in dolomitic marbles, and is progressively succeeded by tremolite, diopside and diopside + forsterite. The conditions at which such reactions take place are however strongly dependent on the composition of the fluid phase present (Yardley, 1989; Chapter 5).

Calc-silicates are very much more variable in their mineralogy, and common phases include actinolite, hornblende, biotite, plagioclase, diopside, microcline, epidote/clinozoisite, zoisite, garnet and sphene. Fluid composition plays a very important role in determining which minerals are developed, in addition to temperature and pressure.

49

Talc marble

Greenschist facies
(Additional examples: **93, 105**)

There are two distinct types of carbonate texture in this rock, fine-grained granoblastic polygonal carbonate in the lower left and coarse porphyroblastic carbonate elsewhere. Talc displays typical second-order birefringence colours, dominantly yellow, and is locally intergrown with calcite. The low relief, low birefringence material in the upper right is albite.

Locality: Campolungo Pass, Ticino, Switzerland. Magnification: × *9, PPL and XPL.*

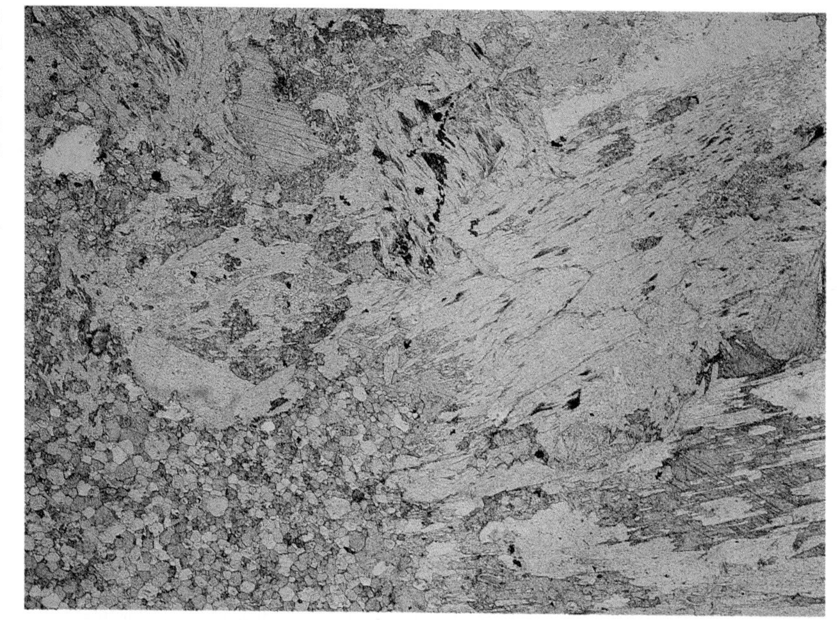

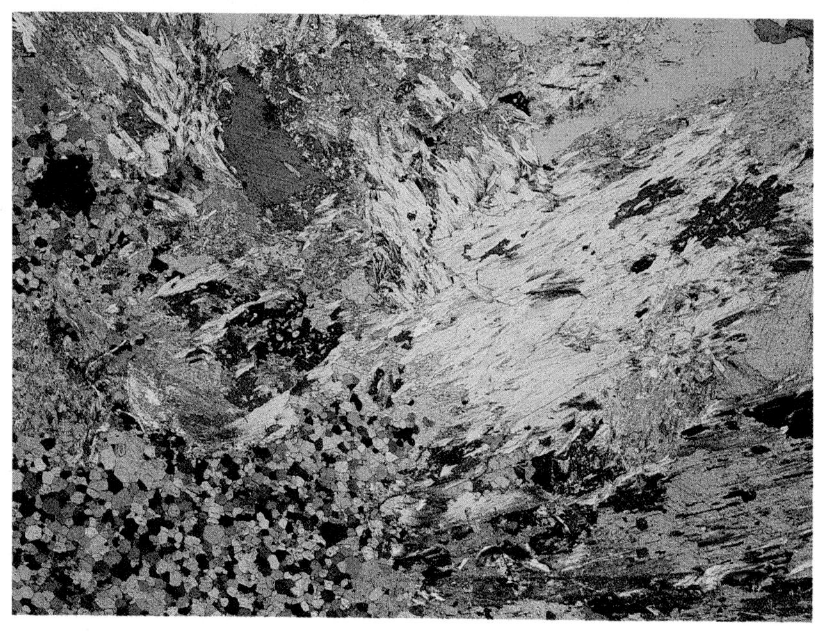

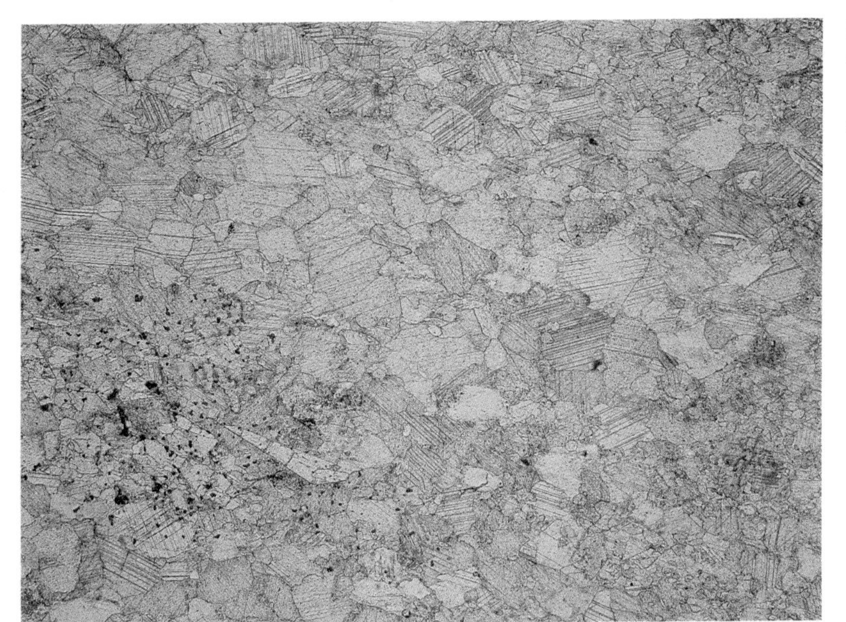

50

Tremolite marble

Amphibolite facies

The main minerals in this rock are calcite, dolomite and tremolite. The tremolite is identified by first- and second-order interference colours and by diamond or wedge shaped crystals; the section is slightly thinner than a standard section.

Calcite and dolomite can in principle be distinguished by the fact that calcite has twin lamellae parallel to the rhomb edges or parallel to the long diagonal whereas in dolomite the twin lamellae lie parallel to the short diagonal of the diamond made by the rhombohedral cleavages.

Locality: Gastacher Wände, east side Dorfertal, Ost-Tirol, Austria. Magnification: × 11, PPL and XPL; and × 16, XPL.

51

Diopside phlogopite marble

Amphibolite facies

This is a foliated marble with a fabric defined by aligned plates of brown phlogopite and by alternating carbonate and silicate layers. The carbonate mineral is calcite. Diopside forms rounded grains which, in this anomalously thin section, exhibit mostly first-order birefringence colours. Higher relief clear clinozoisite is also present and occurs as skeletal grains, as in the lower right corner. Some grains show anomalous blue birefringence. Low relief and birefringence material includes both quartz and plagioclase feldspar. There is a small amount of accessory sphene.

Location: 8 km south of Majavatn, Nordland–Trondelag border, Norway. Magnification: × 14, PPL and XPL.

52

Clinohumite forsterite spinel marble

Pyroxene hornfels facies

Pale yellow clinohumite crystals can be seen clearly in the centre of the field of view. Between two yellow crystals, the small high relief crystals are of spinel and a high concentration of spinel crystals also appears at the bottom left of the field. Both calcite and dolomite are present in this rock and can be difficult to distinguish in the PPL photograph from forsterite. The birefringence of the olivine does however make it clearly visible in XPL and a number of crystals are visible in the lower right quadrant.

Locality: aureole of Bergell tonalite, Val Sissone, northern Italy. Magnification: × 20, PPL and XPL.

53
Scapolite marble
Amphibolite facies

Like many calc-schists this one has an extremely complex assemblage. Sphene is readily apparent in PPL from its high relief and lozenge shape, while hornblende is distinctly pale green. Corroded diopside grains have high relief and bright first-order birefringence colours; scapolite is easily distinguished from diopside by lower relief and alteration rims of plagioclase. It occurs in the centre of the field of view and is shown in more detail in the high magnification image. Clinozoisite displays characteristic anomalous blue birefringence, and is present near the centre of the lower edge of the low power view. Calcite is a common constituent of the matrix, but low birefringence clear quartz and weakly clouded microcline are also common. Microcline is well developed in the upper left corner.

Locality: Deeside limestone, Ord, Banchory, Aberdeenshire, Scotland. Magnification: × 16, PPL and XPL; and × 43, XPL.

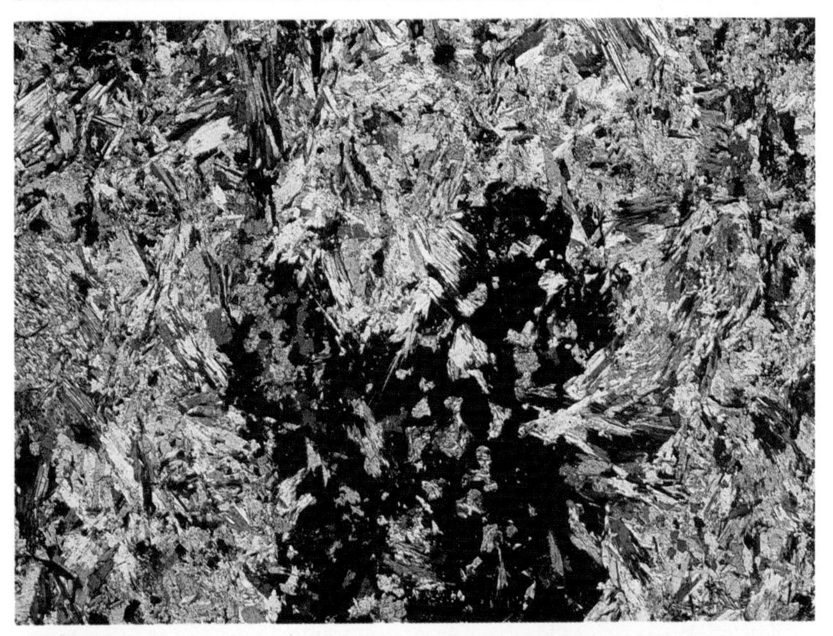

54

Wollastonite diopside grossular calc-silicate rock

Hornblende hornfels facies

These photographs show a large crystal of grossular occupying a major part of the lower half of the field of view. The elongate grains showing grey, black and white interference colours are wollastonite and occupy most of the rest of the field of view. In the higher magnification view more detail of wollastonite can be seen. The mineral showing bright first- and second-order colours is mainly diopside. No carbonate mineral remains.

Locality: Pollagach Burn, Ballater, Grampian Region, Scotland. Magnification: × 7, PPL and XPL; and × 20, XPL.

55

Actinolite andesine schist

Amphibolite facies

In this calc-silicate schist no carbonate remains, but the presence of abundant Ca-silicates attests to the former presence of carbonate in the parent sediment. Colourless, moderate relief actinolite has bright first-order birefringence colours and is often poikiloblastic. Andesine can also have a porphyroblastic habit, but within this field of view is finer-grained. The next most abundant phase is colourless chlorite with distinctive low birefringence, and quartz is also common in the matrix. Minor constituents include some brown biotite and a dusting of opaque graphite.

Locality: Carn Dubh, 9.5 km south of Braemar, Scotland. Magnification: × 32, PPL and XPL.

56

Clinozoisite schist

Amphibolite facies

The dominant mineral in this calc-silicate rock is clinozoisite and the photographs show large and smaller porphyroblasts with zoning apparent from the birefringence. Some zones display characteristic anomalous birefringence but the section is a little too thick. The finer-grained part of the rock contains clinozoisite, blue-green amphibole, brown biotite, pale chlorite, oligoclase and quartz. Brown biotite and green hornblende are present throughout the rock but pale green chlorite and porphyroblastic plagioclase occur in the upper left and lower right corners.

Locality: Lokovista River, Central Rhodopes, Bulgaria.
Magnification: × 11, PPL and XPL.

Metamorphism of igneous rocks

The most widespread meta-igneous rocks are metamorphosed basaltic flows and related minor intrusions, which are prevalent in many metamorphosed sedimentary successions. Acid to intermediate metavolcanics also occur and are common in some terranes, while metamorphosed granites are mostly found in complex polymetamorphic terranes, such as collision zones.

A major distinction between the metamorphism of igneous rocks and sedimentary rocks is that the early stages require addition of water to hydrate (and often carbonate) primary igneous minerals to low grade metamorphic minerals. Only then can subsequent heating lead to progressive devolatilization in an analogous manner to pelites. In the case of thick, massive sills or flows, relic igneous minerals commonly persist up to greenschist facies conditions, and exceptionally beyond. Such behaviour is of course in marked contrast to that of permeable tuffs, illustrated previously (p. 41).

The examples illustrated in this chapter are drawn from regionally metamorphosed rocks and are of considerable antiquity. However much metamorphism of igneous rocks is taking place today in geothermal fields around active or recently active volcanoes. Many such rocks have however little chance of being preserved on a geological time scale.

Metamorphism of basic and intermediate igneous rocks

An essential mineral in metabasites over most of the *P–T* spectrum of metamorphism is amphibole, and indeed many metabasites are known as amphibolites. In marked contrast to pelites, in which distinct zones are defined by the appearance of new phases, reactions in metabasites are often continuous, leading to progressive change in amphibole composition with pressure or temperature. Hence individual pelite zones can rarely be distinguished in metabasites. Instead, the broader, gradational changes in metabasite mineralogy are the basis for the facies classification outlined in the Introduction (p. 3).

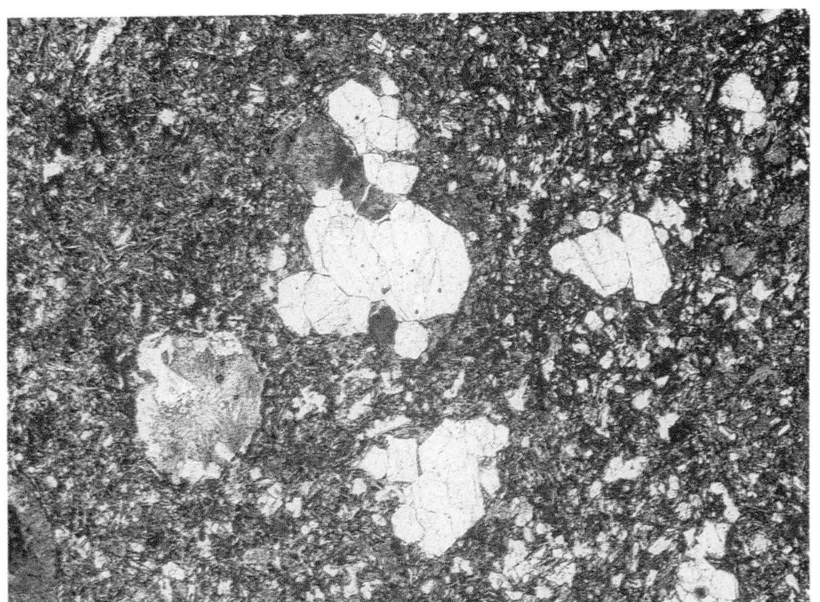

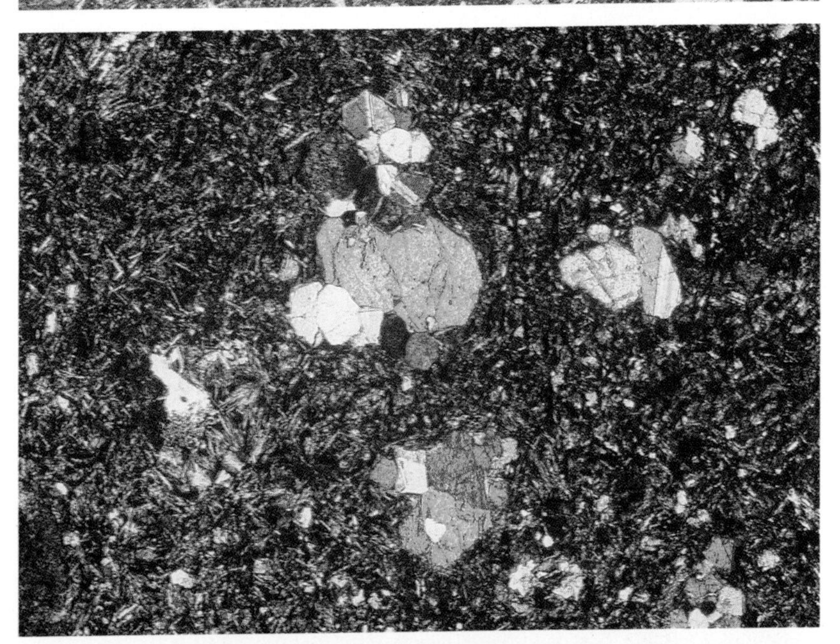

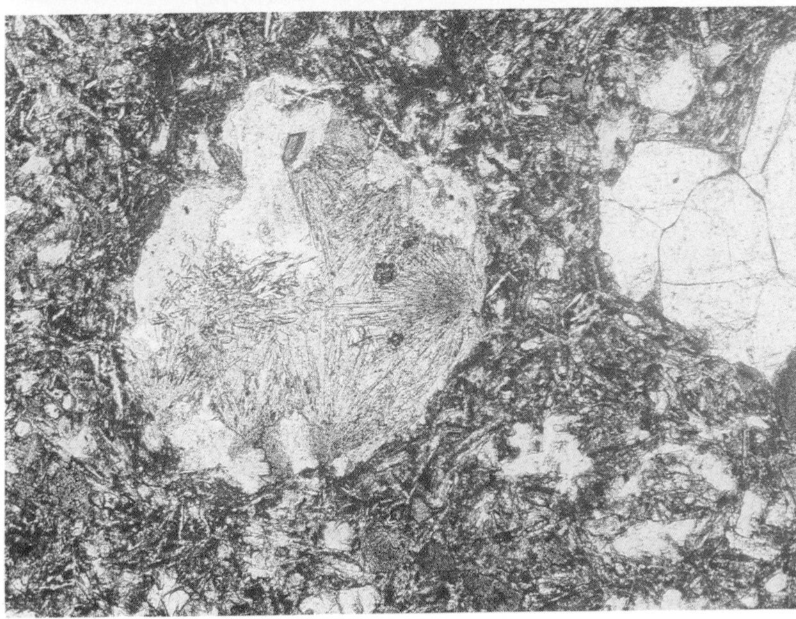

57

Pumpellyite metabasalt

Prehnite pumpellyite facies

This rock is a metamorphosed amygdaloidal pillow lava. The upper photographs show phenocrysts of pyroxene which have been unchanged by metamorphism; the groundmass still contains pyroxene and plagioclase and shows variolitic texture in parts. Interstices between pyroxene and plagioclase are composed of serpentine and chlorite. The enlarged view shows a detail of one amygdale in which radiating bundles of green acicular crystals of pumpellyite, showing green to yellowish-brown pleochroism (compare the two PPL views), make up a major part of the infilling. The colours shown are very characteristic of pumpellyite in metabasalt.

Locality: south of Riverton, South Island, New Zealand. Magnification: × 15, PPL and XPL; and × 40, PPL.

58

Greenschist with igneous relics

Greenschist facies
(Additional example: **81**)

This rock is a low grade metamorphic rock which retains relics of pre-existing igneous minerals and textures, with metamorphic chlorite, epidote, quartz and albite. In the low power view taken under XPL remnants of ophitic texture are fairly clear but metamorphic veinlets of quartz are also apparent.

The detailed high power views show corroded relics of pale brown primary augite, distinguished by their bright birefringence. The fine-grained matrix of the rock contains yellow epidote, chlorite and albite. Small amounts of actinolite are also present and are best seen fringing an augite relic in the upper left quadrant, where they project into an adjacent quartz vein.

Locality: Lake Wakakipu, South Island, New Zealand.
Magnification: × 16, XPL; and × 72, PPL and XPL.

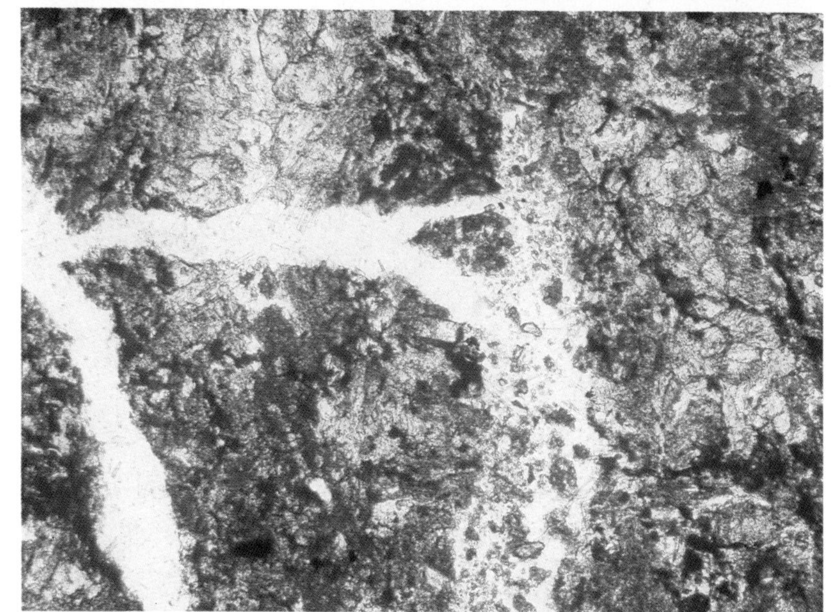

59

Epidote actinolite schist

Greenschist facies

The two upper photographs are low power views which show that this rock is segregated into layers of dark and light minerals. The dark, schistose regions are mainly composed of matted actinolite and chlorite with granules of epidote and accessory sphene. Some muscovite is also present. The light areas are made up mainly of porphyroblastic albite, with some calcite and quartz. The enlarged region shows more detail of the albite crystals and their twinning, together with the matted phyllosilicates and actinolite in the schistosity. This strongly schistose rock with its relatively low content of ferromagnesian minerals may be a metamorphosed volcanogenic sediment or a lava flow.

Locality: near Arrowtown, South Island, New Zealand.
Magnification: × 16, PPL and XPL; and × 28, XPL.

60

Epidote amphibolite

Amphibolite facies

This rock consists mainly of bladed hornblende, greenish-brown biotite and epidote with lesser amounts of plagioclase and calcite; sphene is an accessory mineral. Aligned biotites give the rock a weak schistosity, which is cut across by many amphibole grains and by zoned epidotes whose garish birefringence colours are distinctive.

Locality: Anagh Head, Co Mayo, Ireland. Magnification: × 22, PPL and XPL.

61

Amphibolite

Amphibolite facies

These are photographs of the most common metabasic rock type, consisting mainly of plagioclase and hornblende. Whereas hornblende forms relatively small discrete prisms, these are enclosed in poikiloblastic plagioclase. Small grains of opaque ilmenite occur throughout the rock, and other minor phases present are epidote, biotite and chlorite. Epidote occurs in the upper right and lower left corners and is distinguished by its bright birefringence. Green chlorite occurs in the lower right quadrant while brown biotite is seen best in the upper left one.

Locality: south of Bunaveela, Co Mayo, Ireland. Magnification: × 27, PPL and XPL.

62

Cordierite anthophyllite gneiss

Amphibolite facies

This rock is mainly composed of colourless cordierite, greenish-brown anthophyllite and minor iron oxide. Many of the features by which cordierite can be identified optically are missing in this sample but the birefringence and multiple twinning are characteristic. The anthophyllite is only slightly pleochroic and shows straight extinction.

The Orijarvi locality is a classic one for cordierite–anthophyllite rocks. Their origin was controversial for many years because in composition they do not correspond to any igneous or sedimentary precursor, however they are now believed to form by high grade metamorphism of hydrothermally altered basalt in most instances.

Locality: Orijarvi, Finland. Magnification: × 7, PPL and XPL.

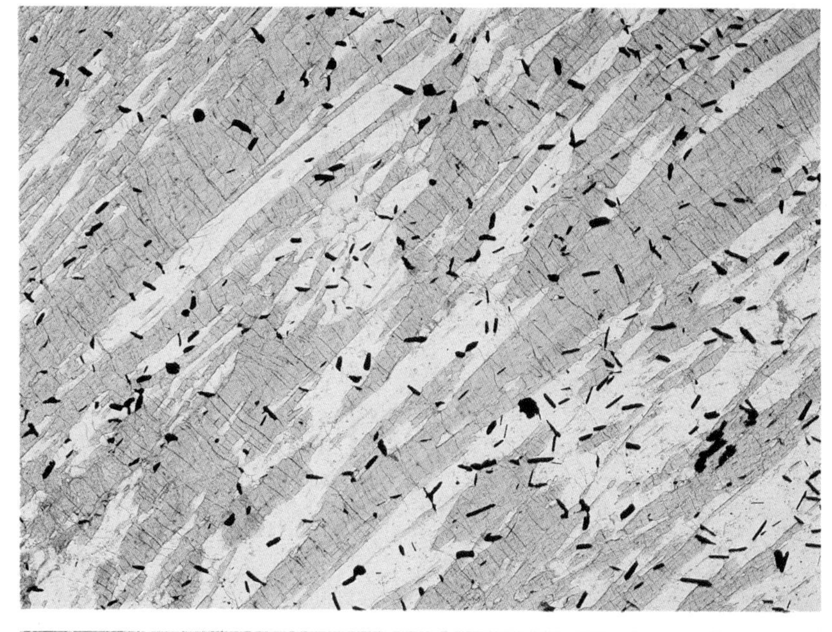

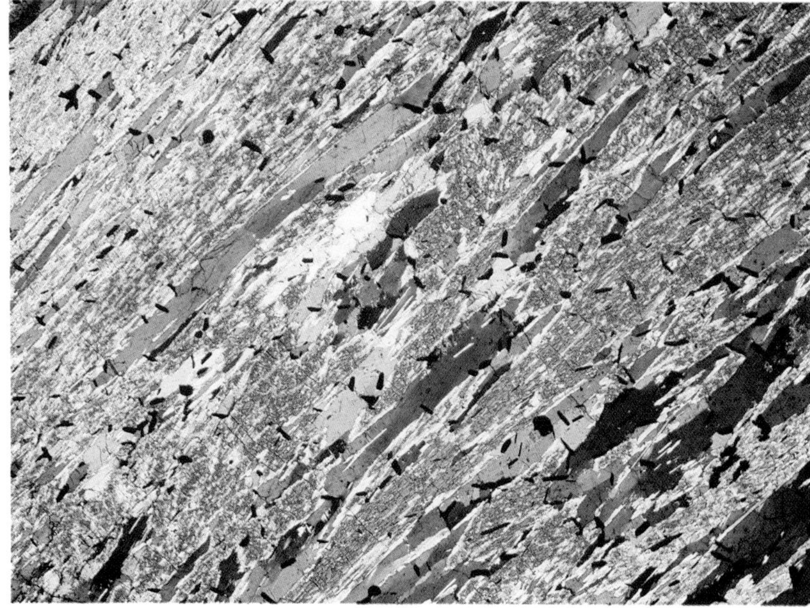

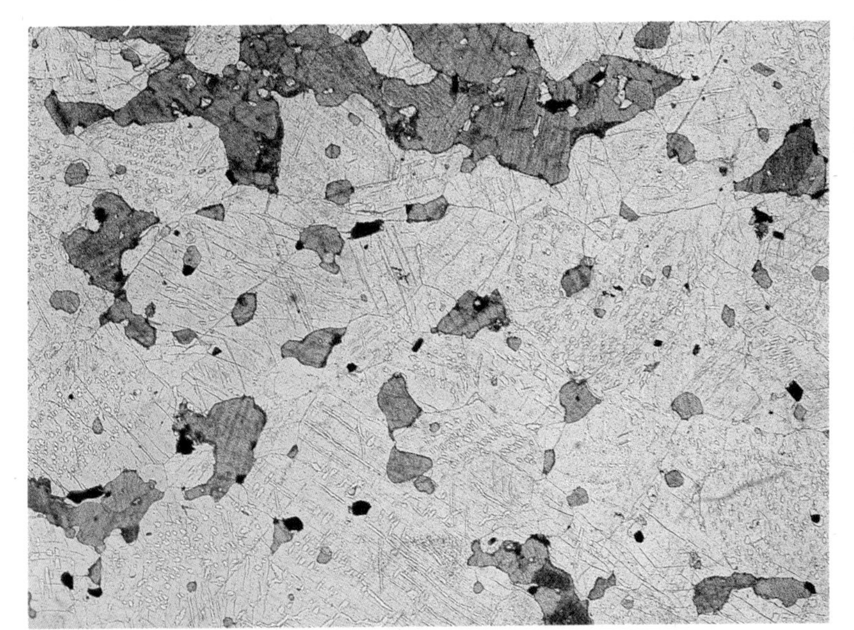

63

Feldspathic granulite

Granulite facies
(Additional example: **80**)

The field of view shown here is occupied by clinopyroxene, orthopyroxene and antiperthitic feldspar. Both pyroxenes have a greenish colour so that they cannot be readily distinguished in the PPL view but under XPL the higher birefringence of the clinopyroxene is noticeable. In the PPL view a slight mis-setting of the focus allows the Becke line between K-feldspar lamellae and blebs in the antiperthite to show clearly against the host plagioclase. Both pyroxenes show exsolution lamellae.

The high magnification view of a region to the right-of-centre of the lower power view provides more detail of the antiperthitic feldspar.

Locality: Scourie, northwest Highlands, Scotland. Magnification: × 8, PPL and XPL; and × 22, XPL.

64

Garnet hornblende pyroxene granulite

Granulite facies

This granulite facies metabasite is of rather uniform grain size and is composed of brown hornblende, garnet and pale green clinopyroxene set in a matrix of plagioclase. In a number of places hornblende appears to replace the pyroxene. This rock is a relatively high pressure granulite, as evidenced by its association with kyanite granulites (**38**); the absence of orthopyroxene is a common feature of such metabasites (cf. **80**).

Locality: Slishwood, Co Sligo, Ireland. Magnification: × 27, PPL and XPL.

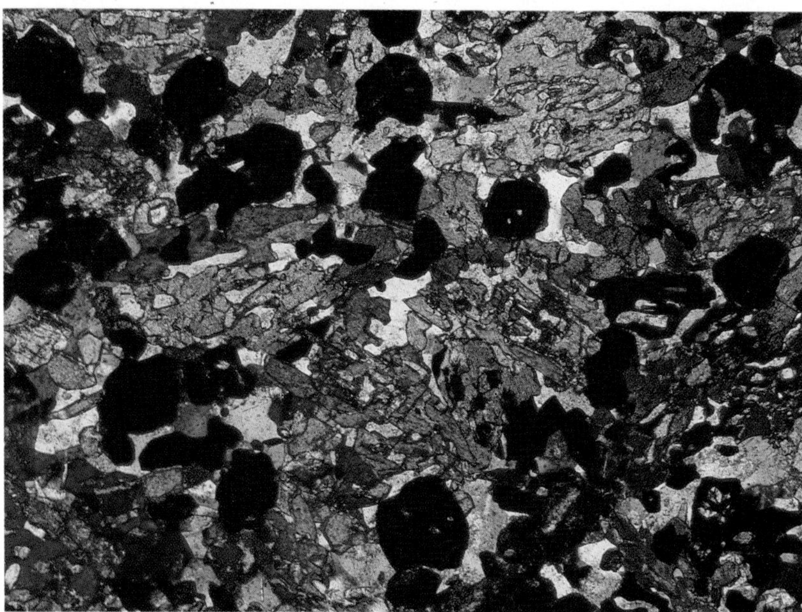

65

Crossite schist

Blueschist facies

This pair of photographs shows the margin of a coarse crossite-rich band in fine-grained crossite schist. The rock consists mainly of strongly pleochroic crossite, with minor epidote (seen in XPL) and apple-green chlorite. The crossite has pleochroism straw-yellow, blue and lavender – this latter colour does not reproduce as well as we would have wished. Zoning is apparent in some grains.

Locality: Shuksan suite, North Cascades, Washington, USA. Magnification: × 28, PPL and XPL.

66

Lawsonite blueschist

Blueschist facies

This rock has the classic blueschist assemblage of lawsonite + glaucophane. In the field of view are two large poikiloblasts of lawsonite, the lower of which displays weak multiple twinning.

Much of the rest of the field of view is composed of fine-grained glaucophane but it has a very pale colour in this rock. The high relief mineral easily seen in PPL is epidote, and other minerals present are chlorite (a coarse seam abuts the left edge below midpoint), muscovite (e.g. impinging on lawsonite in the top left corner), aragonite and a small amount of zircon (not visible here).

Locality: Franciscan Formation, California, USA. Magnification: × 23, PPL and XPL.

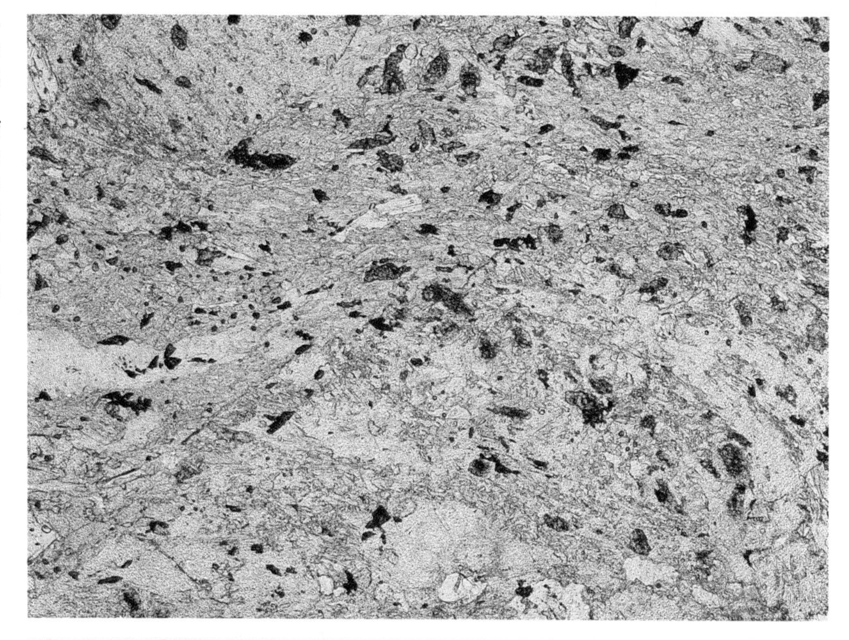

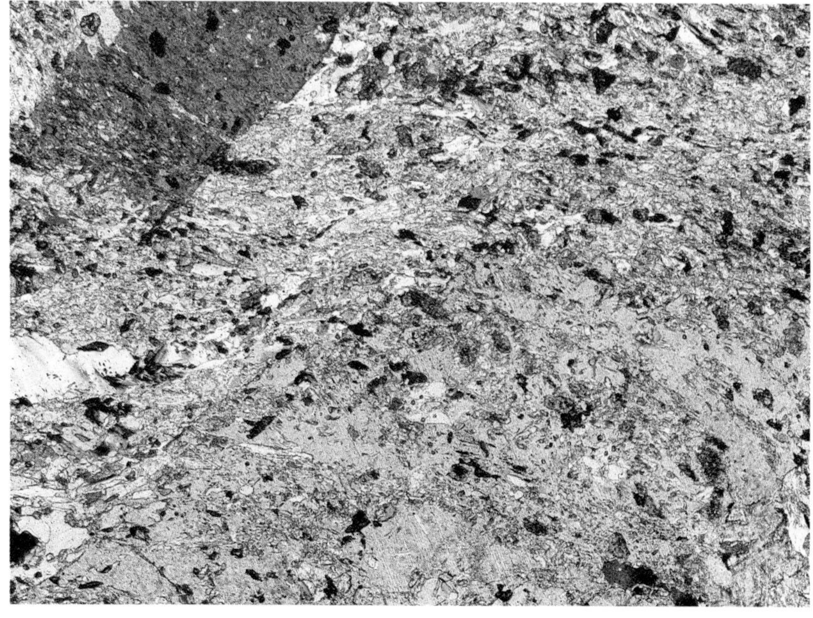

67

Garnet glaucophane schist

Blueschist facies

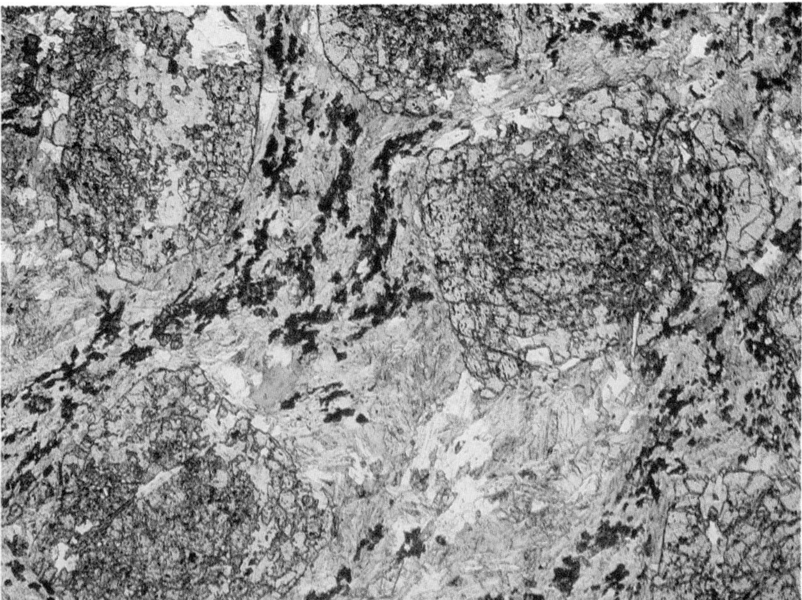

The main minerals in this rock are glaucophane and garnet, the latter being highly altered to chlorite. The small high RI crystals that appear nearly opaque and are interspersed with the glaucophane are of rutile rimmed by sphene: some muscovite is also present.

Many garnets in this rock have cores that are crowded with small inclusions, whereas the rims are relatively free of inclusions. The cores appear to be preferentially replaced by chlorite in some cases, leading to a poorly developed atoll structure (cf. **99**).

This rock may have formed in the eclogite facies and been subsequently recrystallized in the blueschist facies so that original omphacite was entirely replaced by glaucophane (*see also* **107**).

Locality: Franciscan Formation, California, USA. Magnification: × 11, PPL and XPL.

68

Eclogite

Eclogite facies
(Additional examples: **102, 103, 104**)

This rock contains abundant garnet, with pale green omphacitic pyroxene and platy muscovite defining a marked foliation. A small amount of blue glaucophane is also present and there is minor quartz and accessory rutile.

Locality: Kvineset, West Norway. Magnification: × 27, PPL and XPL.

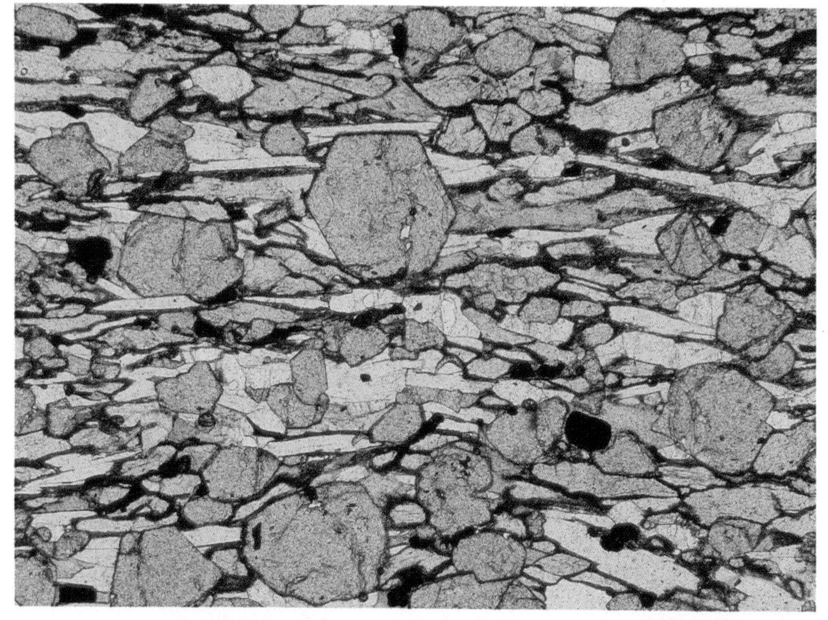

69
Kyanite eclogite
Eclogite facies

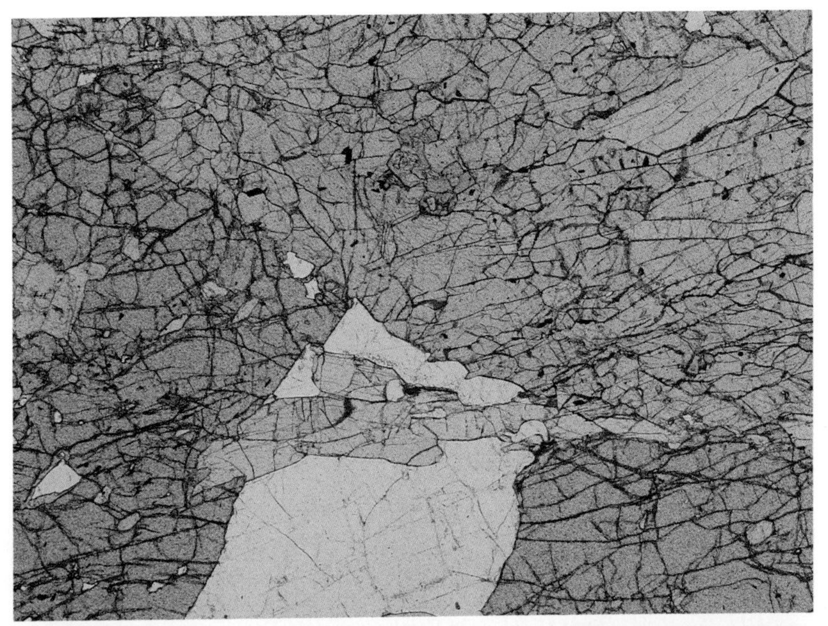

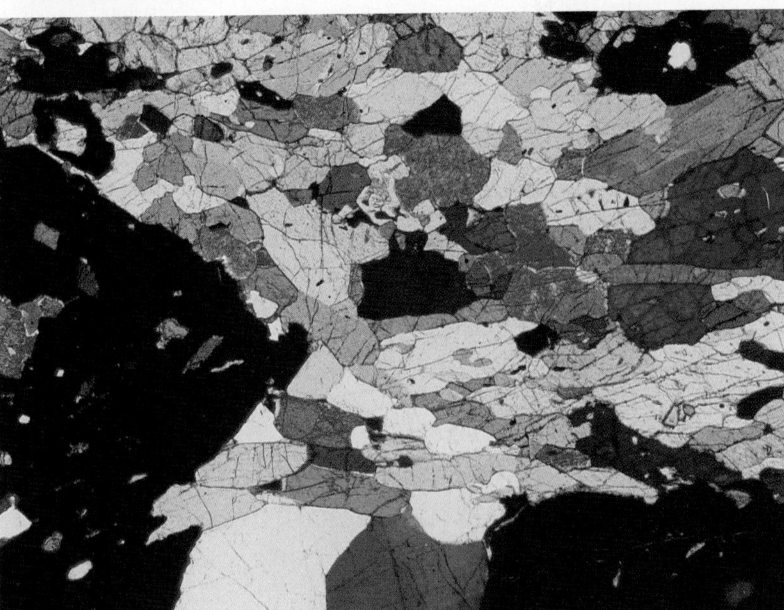

In addition to the essential minerals of an eclogite *viz* an omphacitic pyroxene and garnet, this rock contains kyanite, zoisite and quartz. Garnet, omphacite and kyanite all have very similar relief. That of zoisite is slightly lower (see for example the top right corner). In this specimen the kyanite is unusual in that its blue absorption colour is strong enough to be seen in thin section. Above the centre of the field of view is a cluster of small, intergrown kyanite crystals with diamond shaped sections.

A few inclusions of greenish-brown amphibole can be seen in garnet, along with quartz. Numerous small rutile crystals throughout the rock appear opaque at this magnification. The clear crystals at the centre of the lower edge of the field are quartz.

Locality: Verpeneset, Nordfjord, Norway. Magnification: × 12, PPL and XPL.

70

Eclogitized dolerite

Eclogite facies

This was originally a dolerite-textured metabasic body which has been subjected to eclogite facies metamorphism. The unusual texture is the result of remnants of the ophitic texture being preserved, despite the change in the mineralogy of the rock, notably the growth of garnet. The original pyroxenes are highly altered and pseudomorphed by green metamorphic pyroxene, and are full of iron oxide inclusions. Original lath shaped plagioclases have been almost completely replaced and the clear parts of the thin section are now composed mainly of quartz with some feldspar remnants. Both the quartz and feldspar are full of innumerable small inclusions. Some orange-brown biotite can be seen and minor sphene is present.

At intermediate stages in the eclogitization of such rocks, corona textures are often well developed. Some examples are illustrated in specimens **102–104**.

Locality: Flatraket, Nordfjord area, West Norway. Magnification: × 13, PPL and XPL.

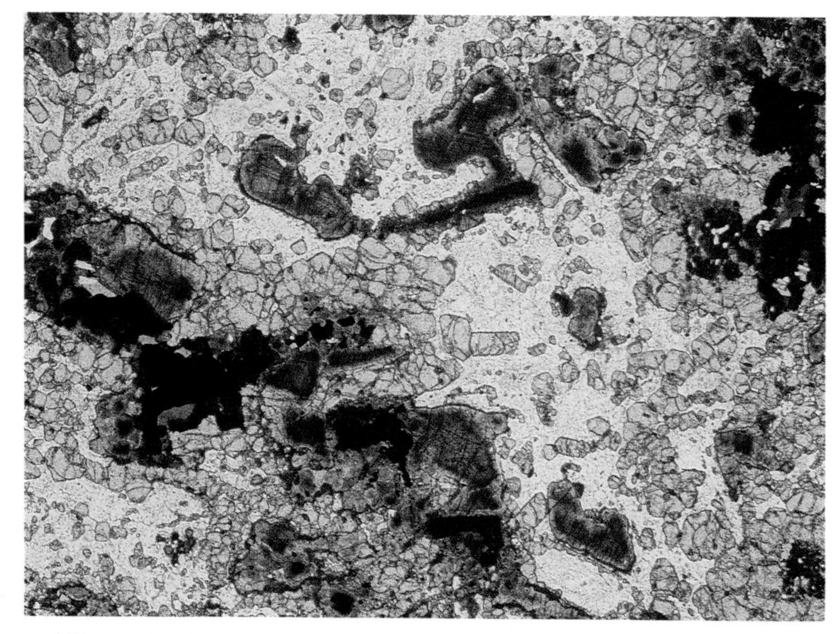

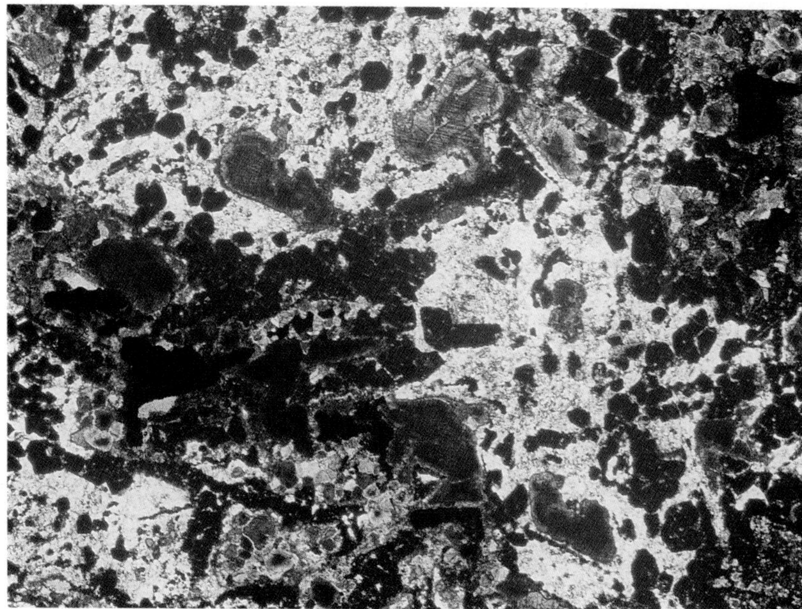

Metamorphism of ultrabasic rocks

Olivine reacts readily with water to produce serpentine, even under low temperature near-surface conditions, and so many ultrabasic rocks have been extensively metamorphosed to serpentinite. This is especially true of *Alpine-type* peridotites, tectonically emplaced in orogenic belts. Serpentinization is usually accompanied by deformation and veining, probably linked to the large volume changes that accompany the process. Some altered peridotites are also carbonated, most commonly magnesite is the carbonate phase.

In some areas, notably where polymetamorphism has occurred, serpentinites may be re-heated and undergo progressive metamorphism. This leads to the re-growth of minerals such as olivine and pyroxene that were present in the original igneous rock, and the resulting rock may be known as a *regenerated peridotite* (**2** is an excellent example).

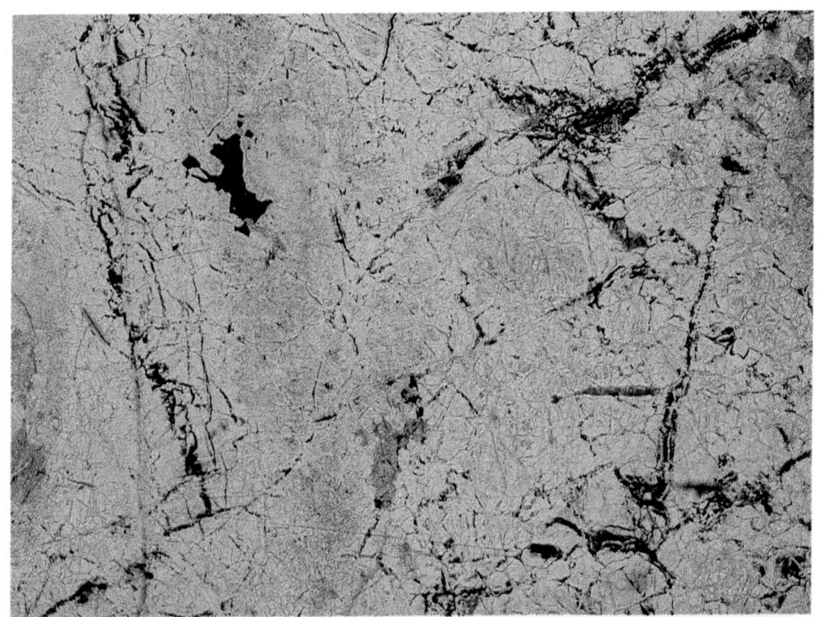

71

Serpentinite

Sub-greenschist facies
(Additional examples: **2, 5**)

This rock is almost entirely made up of serpentine together with opaque iron oxide grains and some brown areas which are possibly hematite stained. Many serpentinites have relics of the original minerals, olivine or pyroxene, within the network of serpentine but no such relic grains are present here. This sample displays a characteristic *mesh texture* in which very low birefringence fine-grained serpentine is divided up into small blocks by numerous thin veinlets of serpentine with slightly higher birefringence.

Locality: Lizard Head, Cornwall. Magnification: × *7, PPL and XPL.*

72

Olivine talc carbonate rock

Amphibolite facies

This is a moderately high grade metamorphosed serpentinite, in which progressive dehydration of the low temperature serpentinite minerals is beginning to restore higher temperature peridotite mineralogy. The large area occupied by talc, showing red and green interference colours, could be misidentified in thin section as a white mica but in hand specimen it is easily identified by being very soft. Olivine occurs as entirely new, metamorphic grains, and the carbonate phase is probably magnesite.

Locality: Valli di Ganano, Val Calanca, Italy. Magnification: × 32, PPL and XPL.

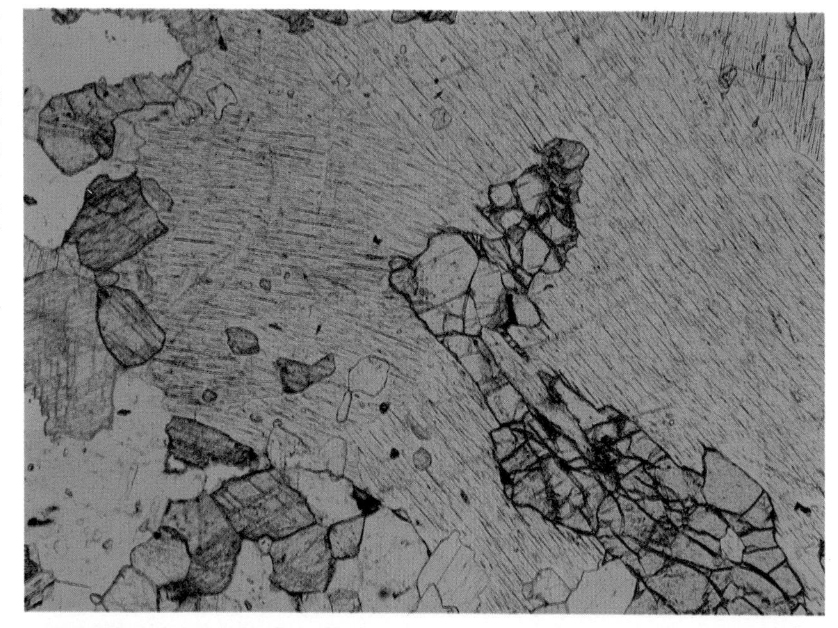

73

Serpentinized meta-peridotite

Granulite facies

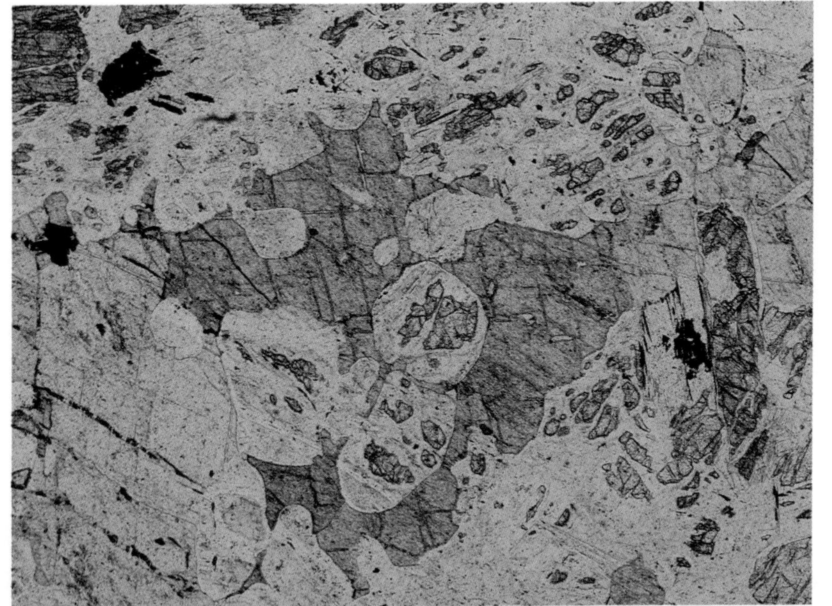

This section is slightly thin since olivine and pyroxene remnants are showing interference colours not higher than second-order.

The centre of the field of view is dominated by poikiloblastic carbonate (magnesite), which encloses a number of distinct grains of olivine, now heavily serpentinized. Other high relief material includes less corroded, low birefringence orthopyroxene, and an elongate enstatite grain lies parallel to the right hand edge of the image. Enstatite, olivine and magnesite comprise the peak metamorphic (granulite facies) assemblage of this rock, and were produced by progressive heating of a previously carbonated and serpentinized peridotite body. Subsequent retrograde hydration has produced chiefly serpentine, but there is a well developed chlorite crystal, next to the enstatite noted above, and minor, highly birefringent talc.

Locality: Ballysadare, Co Sligo, Ireland. Magnification: × 27, PPL and XPL.

Metamorphism of acid plutonic rocks

Since granites and related rocks usually form large intrusive masses which therefore cool very slowly, often accompanied by convective circulation of deep groundwaters, there is a sense in which most granites display metamorphic features simply developed during cooling. Sub-solidus exsolution features abound in alkali feldspars, while both types of feldspar may be heavily altered to clays.

In this section however, we illustrate some examples of acid plutonic rocks which have experienced a distinct metamorphic event. In some examples this metamorphic event has been accompanied, or even initiated, by deformation, while in others the rock retains a granitic texture. More extreme examples of deformed granites are illustrated with the mylonites (see **84–86**).

In very many cases, granites are intruded in the final stages of the orogenic cycle, which is why most are unmetamorphosed. The examples shown here are primarily developed as a result of remobilization of basement rocks.

74

Metatonalite

Greenschist facies
(Additional examples: **85, 86, 87**)

This rock is relatively undeformed and still retains a superficial igneous appearance, primarily because large plagioclase grains retain their original texture. The matrix of the rock is however made up of quartz, biotite and minor K-feldspar which has undergone metamorphic recrystallization. Plagioclase porphyroblasts are distinguished in PPL by a clouding of elongate, randomly oriented microlites. In the enlarged XPL view of the central portion of the lower power image, the microlites can be seen to be crystals of muscovite (typically bright, yellowish birefringence colour) and clinozoisite (anomalous blues). They result from incipient hydration under relatively low grade conditions.

Locality: Obervallach, Austria. Magnification: × 8, PPL; and × 23, XPL.

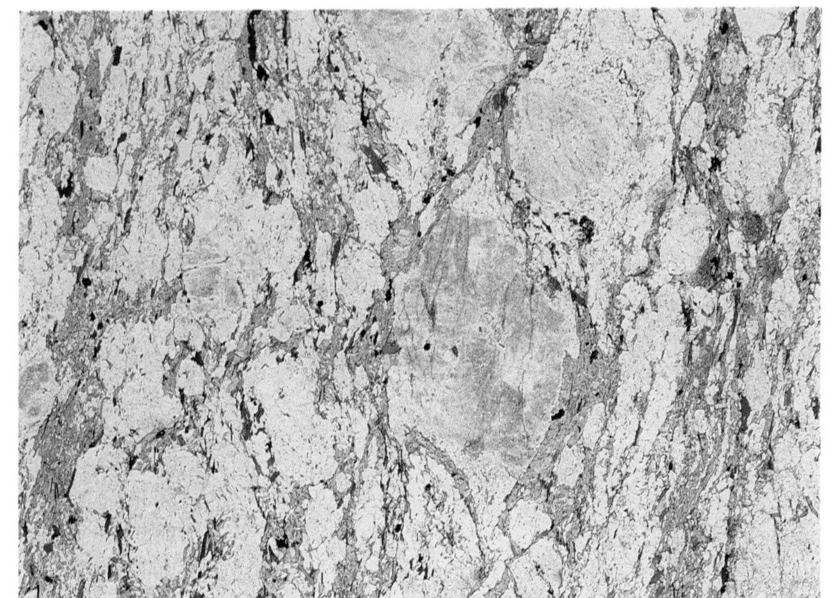

75

Augen gneiss

Amphibolite facies

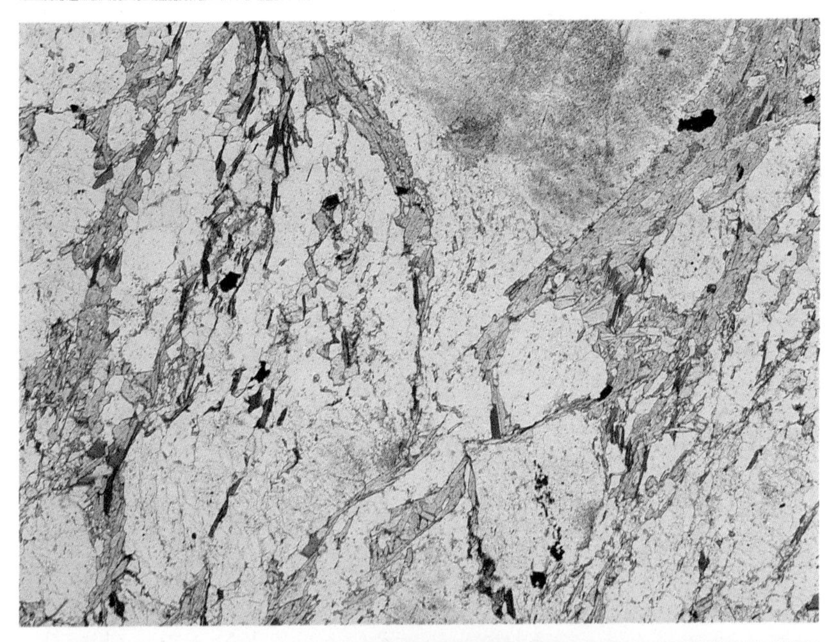

This is a weakly mylonitized rock in which original igneous feldspar now forms porphyroclasts or augen about which the rest of the rock has been deformed. The section has been stained with methylene-blue: this has stained the muscovite in shades of blue. Augen are of both K-feldspar and albite, the matrix of muscovite, biotite and quartz.

In the higher power views it can be seen that part of a large augen at the top is made up of a single crystal of K-feldspar (brown in PPL) with quartz (clear) filling a pressure shadow at the end of the feldspar. The augen composed of albite are clear in PPL also, and are smaller than those of K-feldspar; they tend to be multi-crystalline instead of single crystals. In contrast to the feldspars, quartz has undergone extensive syntectonic recrystallization. It is likely that the muscovite is, in part, of metasomatic origin, formed by fluid influx accompanying the deformation.

Locality: 1 km southwest of Mallnitz, Austria. Magnification: × 8, PPL; and × 20, PPL and XPL.

76

Charnockite

Granulite facies

Charnockite is a granulite facies rock characterized by the coexistence of orthopyroxene and K-feldspar. This rock is rather rich in quartz and contains about equal amounts of sodic plagioclase and microcline perthite, often with distinctive twinning. It is easy to distinguish the K-feldspar from the plagioclase by the low RI of the K-feldspar but the plagioclase is largely untwinned and has RIs very close to those of quartz and so is of oligoclase composition.

The pink to green pleochroism of the hypersthene is very weak, but is sufficient to permit it to be identified as an orthopyroxene. The rock is very fresh except that the orthopyroxene has thin rims of chlorite. The only other minerals present are iron ore and a small amount of biotite.

Locality: southwest of Mount St Thomas, Madras, India. Magnification: × 11, PPL; and × 24, PPL and XPL.

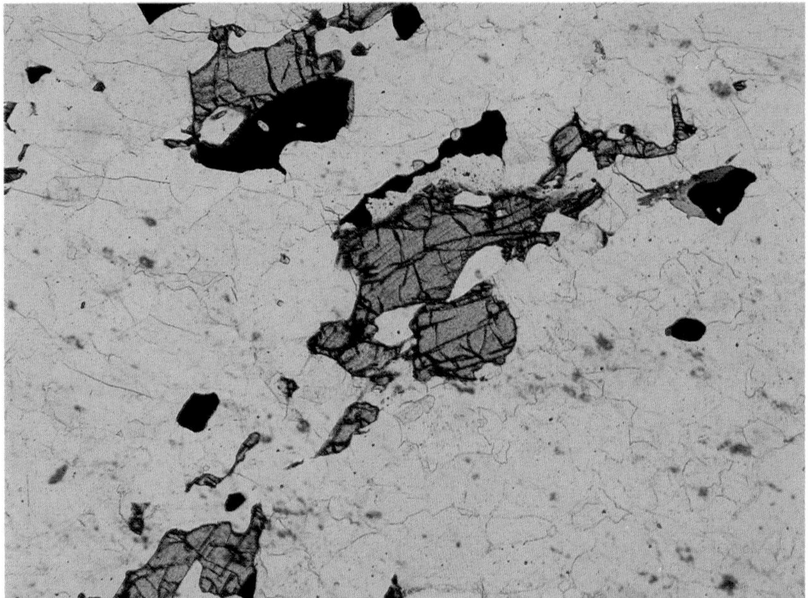

77

Jadeite metagranite

Eclogite facies

In this rock the original plagioclase has broken down under eclogite facies conditions but the granitic texture remains intact. To the left-of-centre of the field is a large mass of fine, high relief jadeite crystals in sheaves; these pseudomorph original plagioclase, and parts of such pseudomorphs are also seen around the edges of the field of view. Primary biotite has partly recrystallized and is now rimmed with small garnets. Phengitic muscovite also occurs in association with the biotite and in the plagioclase pseudomorphs to the right. Original igneous K-feldspar is almost unaffected by metamorphism and a large phenocryst lies along the left edge of the field of view. Primary quartz is also relatively little affected.

Locality: Monte Mucrone, Sesia Zone, Italy. Magnification: × 27, PPL and XPL.
Reference: Compagnoni R, Maffeo B 1973 Schweizerische Mineralogisches und Petrographisches Mitteilungen **53**: 355–82

78

Jadeite gneiss

Eclogite facies

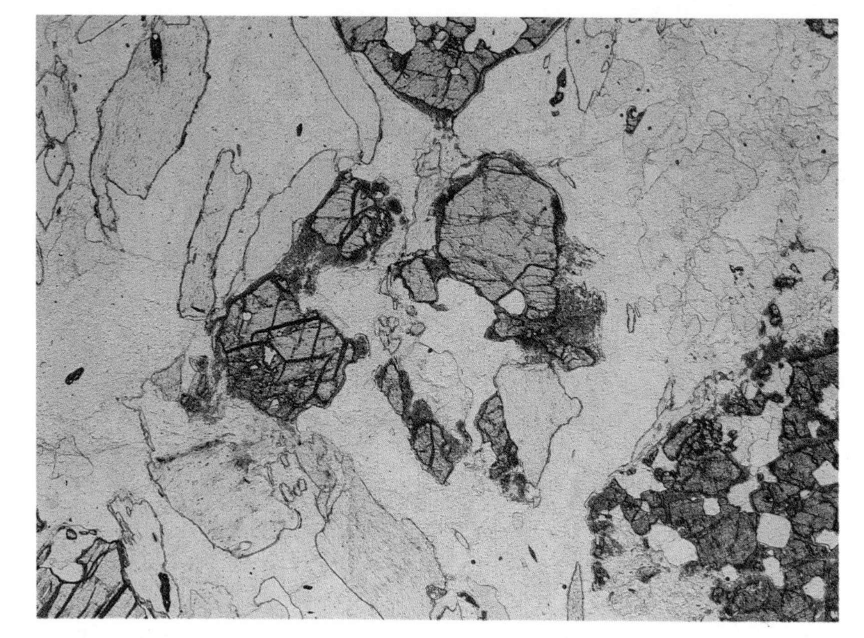

This rock, like number **77**, has been formed from the metamorphism of granite under eclogite facies conditions, but here the metamorphism has been accompanied by deformation. The minerals present are jadeite (both as large high relief crystals and as fine intergrowths with quartz), phengite, microcline and quartz.

The low magnification views show coarse crystals of jadeite, phengitic mica and minor microcline in a matrix of fine, syntectonically recrystallized quartz with some recrystallized microcline.

In the high magnification view the centre of the field is microcline and growing into this crystal from both the left and the right are quartz–jadeite intergrowths which could previously have been a myrmekite of quartz and plagioclase.

The ferromagnesian constituents of the original granite are represented largely by phengite.

Locality: Val d'Aosta, northern Italy. Magnification: × 12, PPL and XPL; and × 52, XPL.

Part 2

Textures of metamorphic rocks

Introduction

The study of metamorphic textures is an essential complement to investigations of their mineral assemblages, because whereas the assemblages help define the physical conditions of the metamorphic event, textures can be indicative of metamorphic processes and the history of metamorphism.

Metamorphic rocks may undergo recrystallization either in response to strain in the absence of any chemical reaction, or as a result of reaction leading to the production of new phases. Often, deformation accompanies metamorphic reactions, in which case the two effects become inseparable.

Particular attention has been paid in many metamorphic studies to the textures of porphyroblastic minerals and their relationship to fabrics in the matrix of the host rock. By identifying the timing of the growth of porphyroblastic metamorphic index minerals, relative to the development of schistosities during deformation, it becomes possible to correlate the attainment of particular metamorphic conditions with specific deformation episodes to which the rock mass has been subjected. A number of examples of such textural relationships have been illustrated.

Sequences of porphyroblast growth and deformation can be deduced even in rocks that have undergone a single cycle of progressive metamorphism. However it is not uncommon for rocks to have undergone two very different metamorphic episodes at different times i.e. polymetamorphism. In the final section a number of examples of polymetamorphic textures are illustrated.

Although the approach here is based on that of Yardley (1989), the reader is also referred to Spry (1969) for a more complete reference for textural terms and to Vernon (1975, 1989) and Barker (1990) for further discussion.

Simple textural terms

There is a very wide range of textural terms for metamorphic rocks available in the literature (*see in particular* Spry, 1969), but only a small number are in universal usage by metamorphic petrologists.

Grain size and shape

Grain growth in metamorphic rocks is influenced by several independent factors. In particular (*i*) the ease of mass transfer through the rock matrix to the sites of growth can affect both grain size and the numbers of inclusions contained; (*ii*) the mechanisms of addition of atoms to grain surfaces can influence grain shape, for example, if atoms are added more readily to a face in a particular orientation, then an elongate or *acicular* crystal may result (**81**); (*iii*) a tendency to minimize surface area can drive recrystallization to more or less equidimensional grain shapes, with planar surfaces, as in *granoblastic polygonal* texture, or *decussate* texture (developed by strongly anisotropic minerals where crystallographic factors compete with surface energy to control grain shape).

As a result, metamorphic minerals may grow as crystals bounded by rational faces (*idioblastic*; **95**, **107**), or have no crystal faces (*xenoblastic*; **24**, **80**) intermediate growth forms are also possible (*sub-idioblastic*). A very important distinction can be made in many rocks, but especially pelitic schists, between relatively fine-grained matrix grains and appreciably larger *porphyroblasts* or *poikiloblasts*. Poikiloblasts (**29**) differ from porphyroblasts in being sieved with inclusions of matrix grains, whereas porphyroblasts have relatively few inclusions (**107**). Usually, the inclusions within porphyroblasts or poikiloblasts are of minerals which also occur in the rock matrix. However, occasionally phases that have entirely reacted out in the bulk of the rock may be preserved as inclusions totally enclosed within porphyroblasts because the included mineral was unable to react with other phases in the rock matrix. Such inclusions are known as *armoured relics* (**96–97**).

While idioblastic grains usually retain the shape in which they grew, the final size and shape of other mineral grains may have changed as a result of recrystallization. In other words, old grains may be replaced by new ones without a change in the modal abundances of the minerals concerned. Recrystallization may proceed by grain boundary migration between adjacent grains of the same phase (Yardley, 1989; p. 154) or by a solution–reprecipitation mechanism.

Foliations

Many metamorphic rocks have been deformed, and this commonly results in the formation of tectonic foliations. The term foliation is a non-genetic one to describe any planar, spaced or pervasive fabric element, be it of primary or metamorphic origin (**79**). Tectonic foliations may result from the alignment of anisotropic grains, or from compositional segregation, normally into leucocratic layers rich in quartz and feldspar and melanocratic layers dominated by micas or amphiboles and other minerals. Grain alignment fabrics are known as *slaty cleavage* where the grain size is very small (**9**), and as *schistosity* when it is coarser (**12, 20**). Most commonly, cleavage and schistosity are planar fabrics and can both be termed S-fabrics. Usually, the aligned minerals are platy phyllosilicates, but they may be tabular grains of quartz or carbonates (grain-flattening fabric) (**79**). Where the aligned mineral grains are prismatic rather than platy (notably amphiboles), the fabric may be dominated by a strong linear alignment of the grains (lineation) and may be termed an L-fabric (**62**).

Tectonic foliations are commonly refolded by subsequent deformation, producing microfolds or *crenulations* of the earlier foliation. The development of crenulations is often accompanied by a degree of metamorphic segregation (Yardley, 1989; p. 169), with quartz becoming concentrated in the hinges of microfolds and phyllosilicate minerals in the limbs (**11**, **92**). While in some examples (e.g. **10**, **11**) crenulation fabrics are clearly apparent, in others (e.g. **89**) the second deformation has been so intense that the early fabric is largely obliterated.

Where an intense deformation postdates the formation of porphyroblasts, the relatively rigid porphyroblast may shield the material around it from the effects of strain in a plane perpendicular to the principal compression, whilst causing more intense strain around the parts of the porphyroblast at which the principal compressive stress is directed. The resulting heterogeneous strain may lead to localized pressure shadows around the porphyroblast (**83**) and also to deformation partitioning (Bell, 1981) whereby the rock is divided into less strained planar domains containing the porphyroblasts, separated by more intensely strained zones in which the new fabric is best developed.

79 Foliations (bedding and schistosity) in quartzite

The original bedding in this impure quartzite is clearly picked out by a layer rich in opaque grains which corresponds to a heavy mineral enriched layer in the parental sand. The rest of the rock is dominated by quartz with microcline feldspar (whose incipient alteration gives it a clouded appearance in PPL). Grains of quartz, feldspar and the opaque phase are all elongated and define a diagonal tectonic fabric which is a form of schistosity known as a grain-flattening fabric.

Locality: Achill Island, Ireland. Magnification: × 12, PPL and XPL.

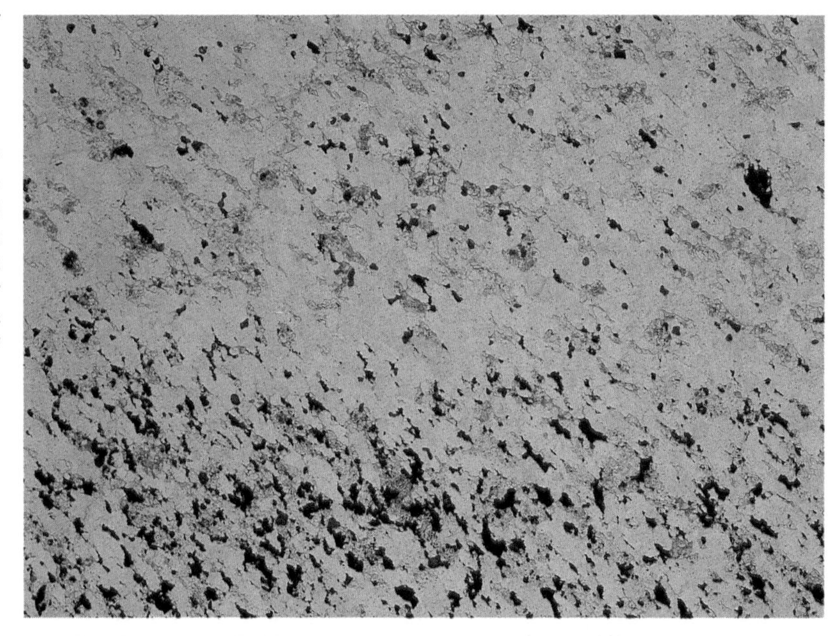

80 Granoblastic polygonal texture in hornblende scapolite granulite

The dominant constituent of this rock is hornblende. Hornblende grains have recrystallized to a texture in which the surface area per unit volume is a minimum for the grain size. Grain boundaries are straight and where three grains meet, they tend to do so in a symmetrical way at *triple junctions*, with an interfacial or dihedral angle of about 120°. This is a granoblastic polygonal texture and represents a close approach to textural equilibrium. It is best seen in the granulite facies, except for the case of carbonate or quartz-rich rocks.

Scapolite, orthopyroxene ('hypersthene') and an opaque phase occur as smaller grains, often isolated at triple junctions. Scapolite is clear in PPL, but in this sample shows a patchy grey appearance caused by the presence of numerous aligned inclusions of needle-like grains of an opaque phase. In XPL, scapolite has a characteristically wide range of bright birefringence colours, ranging through the first- and second-order. Orthopyroxene displays distinctive pleochroism from pink to very pale green; many orthopyroxene grains have undergone incipient retrograde breakdown and are rimmed with colourless amphibole.

Locality: Scourie, northwest Scotland. Magnification: × 16, PPL and XPL.

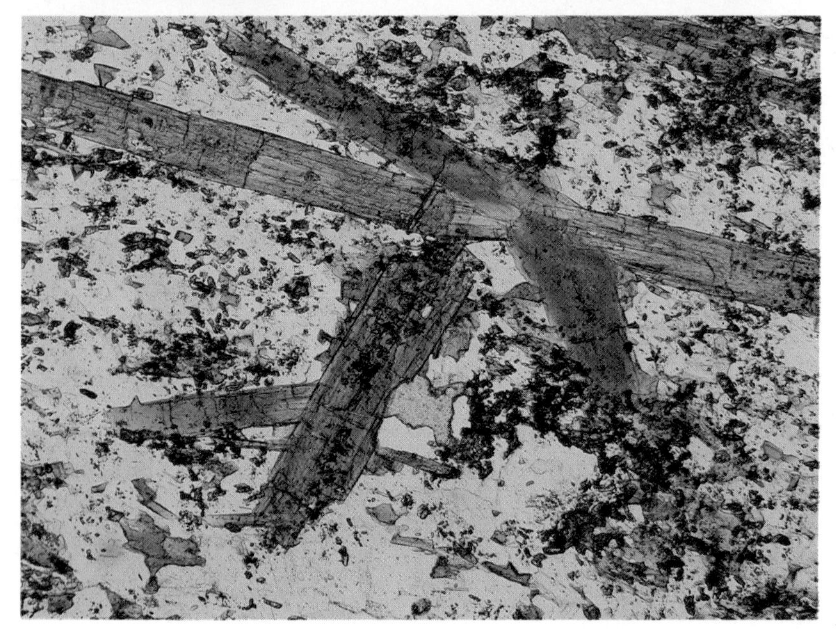

81 Acicular texture in actinolite schist

The elongate, acicular, shape of the large actinolite grains is typical of amphiboles formed at low to medium grades. Their patchy colouration and birefringence is the result of complex compositional zonation. The green colouration of matrix chlorite is very similar to that of the actinolite in PPL, but the birefringence is much lower. The granoblastic matrix of the rock is predominantly of quartz with some albite. A weakly cleaved albite grain is visible near the bottom edge towards the left corner, at the end of an actinolite grain. Minor calcite is also present. Epidote occurs as very small high relief grains, sometimes distinctly yellow in colour with bright birefringence colours; it is present as inclusions in actinolite as well as in the matrix.

Locality: Coronet Peak, Otago, New Zealand (Cambridge University Collection No 39811). Magnification: × 28, PPL and XPL.
Reference: Hutton CO 1938 Mineralogical Magazine **25**: 207–11

Acicular texture (continued)

82 Decussate texture in garnet mica schist

This is a muscovite rich layer of a garnet mica schist and the muscovite displays a decussate texture. This consists of randomly oriented interlocking platy, prismatic or elongated crystals. It differs from granoblastic texture in that the crystals are not equidimensional, but similarly represents an approach towards a minimum surface energy equilibrium texture. Other phases visible include quartz, minor biotite, and heavily altered plagioclase which appears pale brown in the PPL view.

Locality: Beinn Dhubhchraig, near Tyndrum, Tayside Region, Scotland. Magnification: × 15, PPL and XPL.

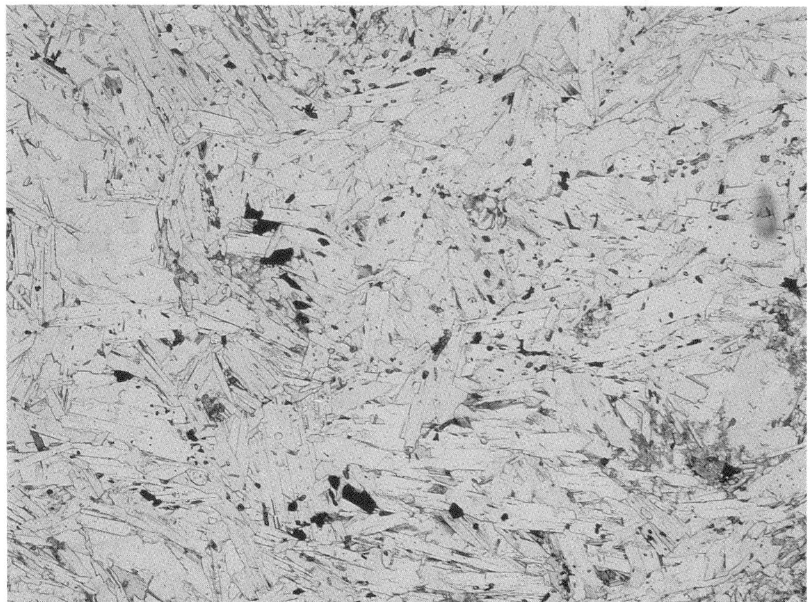

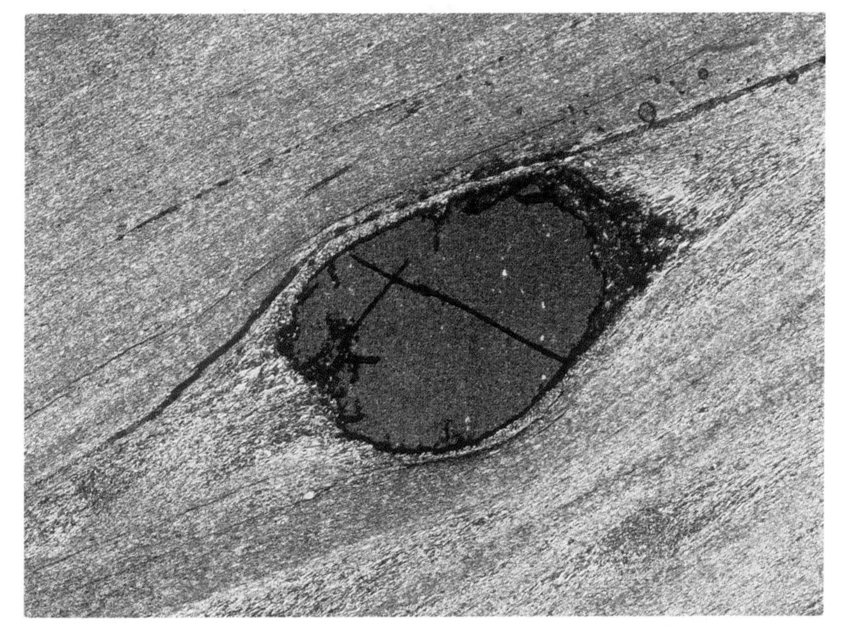

83 Porphyroblasts and pressure shadows in siderite phyllite

The upper photo is a low power view of a very fine grained, low grade (chlorite zone) phyllite in which porphyroblasts of siderite occur. The dominant matrix minerals are muscovite and quartz. It is used here to show the formation of pressure shadows around the porphyroblasts, where quartz has recrystallized into the low strain zone adjacent to the porphyroblast.

The crystal illustrated in the high power view is near the top right corner of the low power view.

Locality: Beavertail, Conanicut Island, Rhode Island, USA. Magnification: × 7, XPL and × 30, XPL.

Plastic deformation and mylonitization

Different minerals respond to deformation in different ways. Quartz and carbonate readily undergo plastic deformation under most crustal metamorphic conditions, as does olivine under mantle conditions. On the other hand many other minerals are relatively rigid and brittle under the same conditions, whilst micas may deform by kinking (**8**). Quartz responds to small strains by the development of sub-grains which may be aligned as parallel deformation bands (**84**). Deformed grains replace one another along sutured grain boundaries or are recrystallized to a fine-grained mortar of new unstrained grains. The process whereby new grains progressively grow, become strained and are replaced is known as syntectonic recrystallization. In extreme cases it can produce a fine, strongly aligned ribbon texture (**88**). Similar textures are produced in olivine-rich rocks at high temperatures (**5**).

More rigid minerals such as feldspar and garnet tend to deform in a brittle manner (cataclasis) and often fail to recrystallize even when their quartz and mica matrix is deforming extensively, but remain as porphyroclasts which may be angular (**75**, **87**) or rounded (**86**, **88**). In the case of feldspar porphyroclasts, the outer portion sometimes recrystallizes to a fine polygonal mortar of sub-grains surrounding a relic core (**38**, **85**).

A rock in which the matrix has undergone extensive syntectonic recrystallization to a finer grain size, leaving large grains remaining as porphyroclasts, is known as a *mylonite*. In *protomylonite* the porphyroclasts still predominate, while in an *ultramylonite* they have largely been eliminated.

In this section a range of rocks illustrating varying degrees of plastic deformation and mylonitization are illustrated while others appear elsewhere (e.g. **5**, **38**, **75**).

84 Strained quartz with sutured boundaries in garnet mica schist

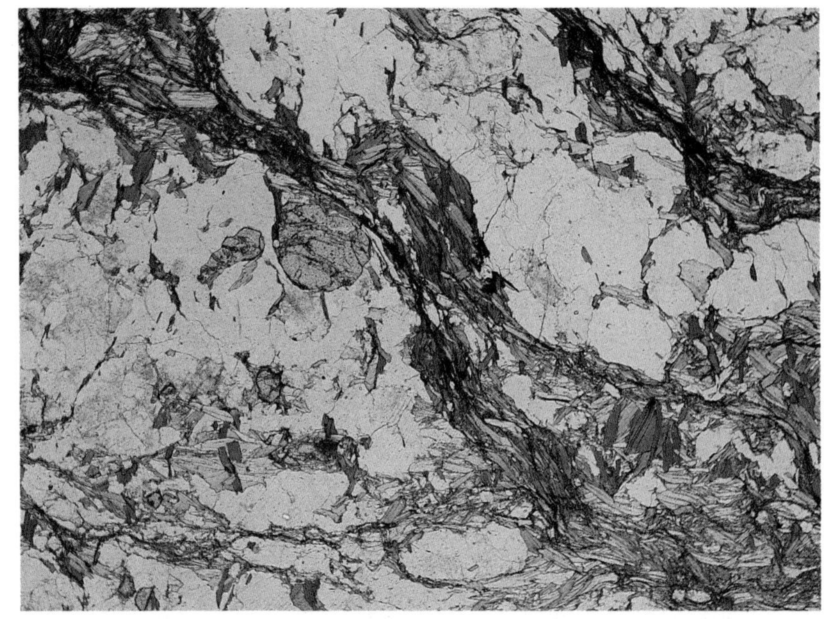

This rock has undergone a pronounced segregation into biotite-rich layers and coarse bands of quartz and plagioclase, and this diagonal segregation fabric has been subsequently flattened to produce a new fabric that is aligned roughly N–S. Garnet grains are small and anhedral; they may have been fragmented, partially broken down or both.

In the leucocratic layers, feldspar is in some cases distinguished by twinning in XPL while in PPL it appears somewhat clouded by incipient alteration. Original large quartz grains are now composed of parallel deformation bands i.e. stripes of slightly different extinction position. Quartz grain boundaries are also highly sutured due to strain-induced grain boundary migration, and in a few areas small undeformed quartz grains have begun to develop along grain boundaries, heralding the onset of syntectonic recrystallization.

Locality: North Cascades highway, Washington, USA. Magnification: × 11, PPL and XPL.

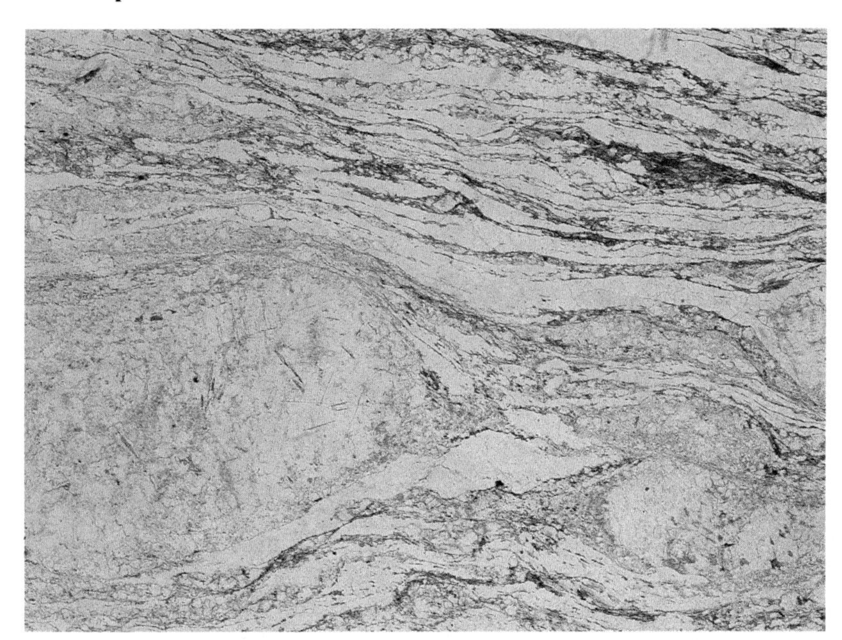

85 Granite mylonite

Intense deformation has disrupted original K-feldspar phenocrysts to produce porphyroclasts with patchy extinction and fracturing; their rims partially recrystallized to a mortar of fine microcline. The remaining texture of the granite has however been almost completely destroyed. Quartz occurs in recrystallized layers separated by segregations of fine biotite and muscovite. Plagioclase is also present in the matrix but is not abundant, and this, with the presence of muscovite, suggests that deformation may have been accompanied by metasomatism with plagioclase breakdown and mica growth (cf. **75**).

Locality: south Brittany, France. Magnification: × 7, PPL and XPL.

86 Granite ultramylonite

Although this sample comes from the same shear zone as the previous one, deformation has been still more intense. K-feldspar porphyroclasts remain, although their margins are strongly fractured. The matrix is composed predominantly of very fine phengitic muscovite giving rise to a distinctive yellow-brown appearance in XPL. This matrix is also dotted with fine quartz. The fine mica probably grew under quite low temperature conditions (greenschist facies), and extensive metasomatism must have accompanied its growth since, for example, plagioclase is absent except at the borders of the K-feldspar porphyroclasts where it sometimes forms a myrmekitic intergrowth with quartz (not readily apparent from these pictures).

Locality: south Brittany, France. Magnification: × 7, PPL and XPL.

87 Ultramylonite

An example of a schist that has undergone extreme plastic deformation.

The PPL view shows a few larger, angular fragments of plagioclase feldspar and an opaque phase occurring as porphyroclasts in a fine-grain matrix of quartz and muscovite with minor sphene and carbonate. The parental rock was amphibolite facies Moine Schist, and even the porphyroclasts are very much smaller than likely parental grains occurring in undeformed schist outside the mylonite zone. Note the later diagonal shears that disrupt the main mylonite fabric.

Locality: Stack of Glencoul, Moine Thrust Zone, northwest Scotland. Magnification: × 40, PPL.

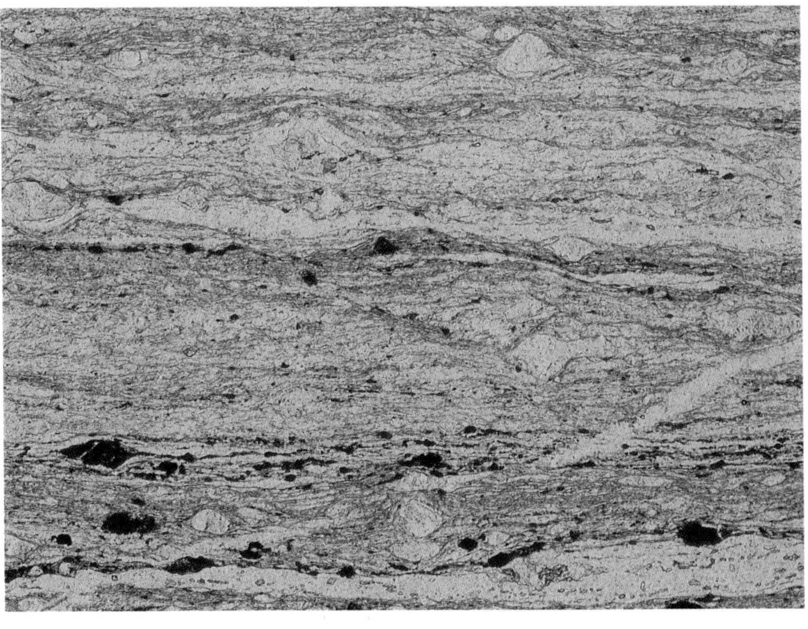

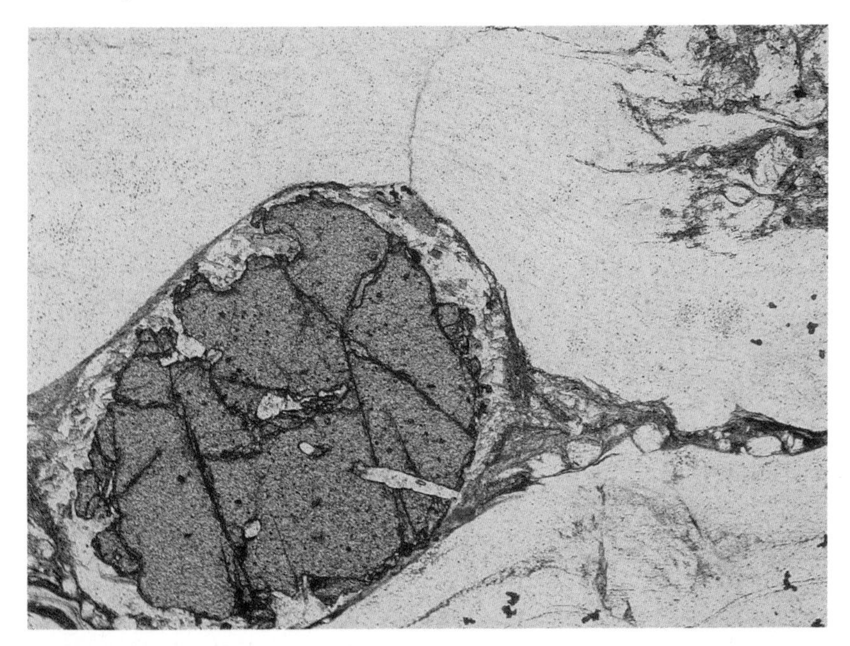

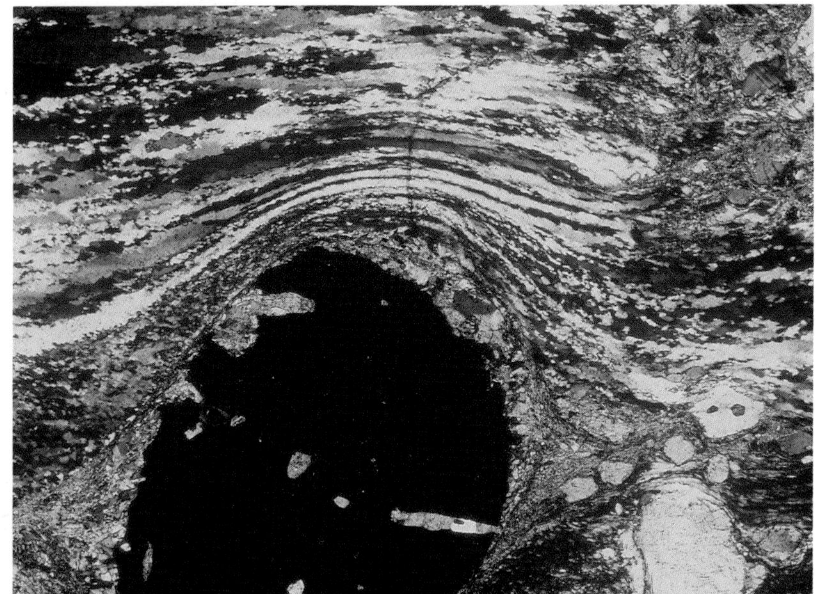

88 Mylonitized garnet mica schist showing porphyroclasts and ribbon texture

Intense strain of a parental garnet mica schist has produced this mylonite. Relatively rigid garnet and plagioclase feldspar survive as abraded relics or porphyroclasts, but quartz has undergone extreme syntectonic recrystallization to produce a fine-grained mortar of recrystallized grains. These show alignment in the areas of most intense strain, around the porphyroclasts, to produce pronounced ribbon texture. Biotite is preserved in strain shadows around porphyroclasts but has been reduced to fine-grained material where deformation was more intense.

Locality: North Cascades highway, Washington, USA. Magnification: × 32, PPL and XPL.

Time relations between deformation and metamorphism

Because porphyroblasts are rigid and do not respond to deformation in the same way as most matrix minerals, it is often possible to use the relationships between matrix foliations and porphyroblast grains to deduce the sequence in which they are formed (Rast, 1958; Voll, 1960; Zwart, 1962; Vernon, 1989).

When porphyroblasts grow they often enclose small grains of matrix minerals, and where these grains had a strongly anisotropic shape, or were concentrated into segregated layers by earlier deformation, the matrix foliation becomes preserved within the porphyroblast and is known as an internal schistosity (S_i). Quartz, ilmenite and graphite are particularly common minerals that define internal schistosities. If no further deformation occurs the internal schistosity of the porphyroblast remains parallel to and continuous with the external schistosity (S_e) of the rock matrix. However if subsequent deformation does take place, modifying the earlier fabric and developing new ones, the porphyroblast will nevertheless preserve the earlier fabric intact. Some of the textures that can result from different time relationships of porphyroblast growth and deformation are portrayed schematically in Figure B.

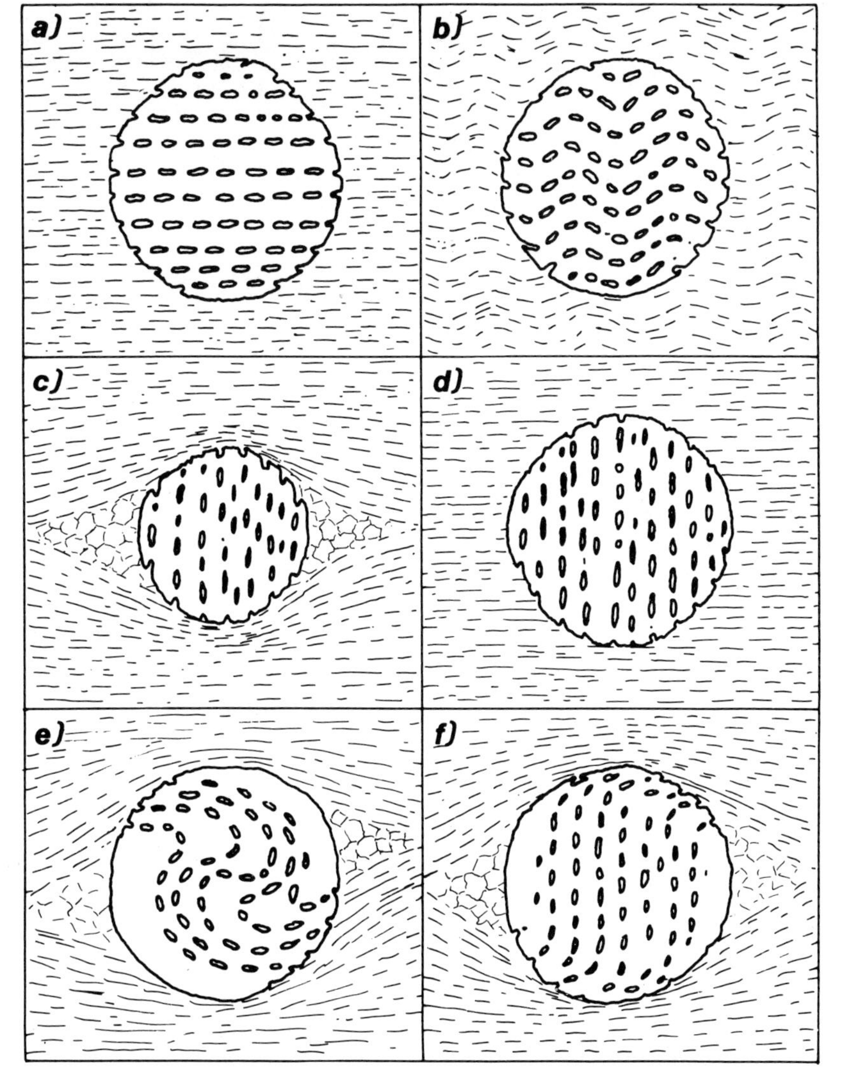

FIG. B Examples of relationships between internal schistosity in porphyroblasts or poikiloblasts, and external schistosity, from Yardley (1989). (a), (b) Examples of post-tectonic porphyroblast growth in which S_i is continuous with and parallel to S_e. (c), (d) Porphyroblasts formed prior to the external schistosity and preserving an internal schistosity that is oblique to it. In (c), a pressure shadow has developed around the porphyroblast, but more rarely, as in (d), there is little disruption of the external schistosity. This is most common where the matrix is almost entirely of mica. (e), (f) Syntectonic porphyroblasts. (e) is a classic snowball *garnet (cf.* **93***) with about 180° rotation during growth while (f) is a more common rotational porphyroblast, often produced by recrystallization of the schistosity into a new orientation during the later stages of its growth.*

Controversy continues to surround the interpretation of some syntectonic porphyroblast textures. Curvature of the inclusion trails within a porphyroblast is usually taken to imply rotation of the porphyroblast relative to the external schistosity as it grew, and hence syntectonic growth. The classic interpretation of such textures is that the porphyroblast was *rolled* by simple shear along the schistosity plane, but this does not appear to apply in all cases. It is apparent in **89** from the near parallelism of the internal schistosities in plagioclase grains with one another and with the crenulation hinge fabric that the porphyroblasts here did not rotate during the development of subsequent foliations. Following an original suggestion by Ramsay (1962), the parallelism of S_i across fold hinges has been demonstrated in field studies by Fyson (1975, 1980), de Wit (1976) and Bell (1985). Thus it may be the external schistosity that progressively rotates relative to the porphyroblast, rather than the reverse, in these examples (see Yardley, 1989, Fig 6.13), and this can account for rotations of S_i of up to 90° if growth was syntectonic. A third explanation for such rotational fabrics (e.g. **90**) is that they simply result from post-tectonic growth over a microfold. In some instances (e.g. **93**) much more extreme rotation of the inclusion trails in porphyroblasts is recorded. These are the *snowball* textures investigated by Rosenfeld (e.g. 1970) and many other workers. Apparent rotations greatly in excess of 90° have been reported and an origin for textures such as **93** by rotation of the porphyroblast seems inevitable. Nevertheless, Bell and Johnson (1989) have recently claimed that this may not necessarily be the case. In one of our examples (**95**) the evidence for rotation of porphyroblasts relative to one another during a later deformation seems unequivocal.

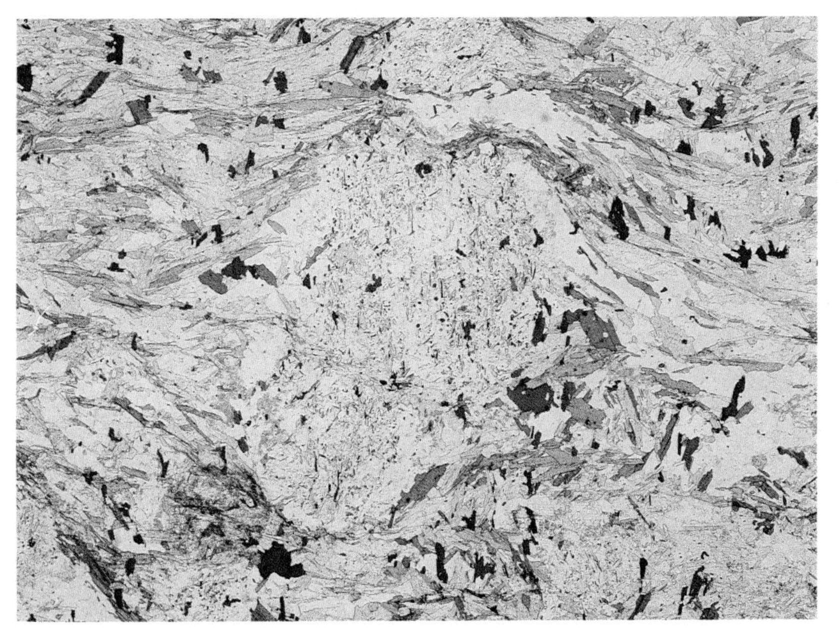

89 Pre-tectonic porphyroblasts in plagioclase biotite schist

The dominant fabric of this rock is a schistosity defined by aligned plates of biotite and muscovite and oriented approximately E–W. In the lower right corner however, this schistosity is less intense and can be seen to have arisen by crenulation of an earlier, approximately N–S fabric. Plagioclase occurs as large porphyroblasts. It is apparent in XPL that some grain rims have a different composition from the cores (top centre and bottom left), due to the overgrowth of oligoclase rims on albite cores. This is quite a widespread phenomenon in garnet zone rocks formed at temperatures where the presence of a *peristerite gap* precludes intermediate compositions between these two end-members. The plagioclase porphyroblasts contain abundant inclusions of small biotite, muscovite and quartz grains, with sparse very small garnet inclusions. Of these, the micas in particular define an internal schistosity that is also aligned N–S in the larger plagioclase grains.

The discordance between internal and external schistosities, and the pronounced flattening of the main, external, schistosity around the porphyroblasts provide firm evidence that plagioclase grew before the deformation that produced the main schistosity of the rock, but after the early deformation that gave rise to the N–S fabric.

Locality: Nephin Beg Range, Mayo, Ireland. Magnification: × 8, PPL and XPL.

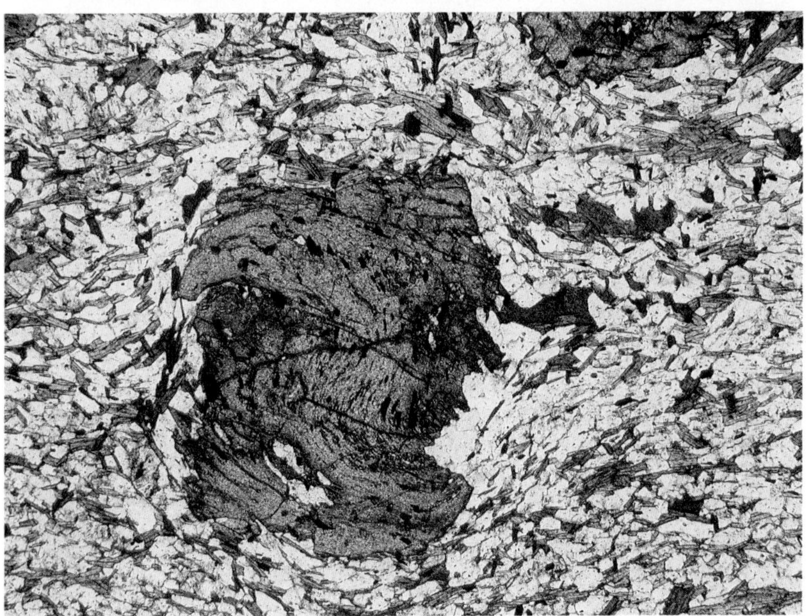

90 Probable syntectonic porphyroblast in garnet biotite schist

The garnet porphyroblast shown here has curved trails of small opaque inclusions of ilmenite which apparently indicate about 90° of rotation of the garnet during growth. There are three ways in which this pattern could have arisen: (1) post-tectonic growth of garnet over a pre-existing crenulation cleavage, destroyed by subsequent deformation in the rest of the rock; (2) syntectonic garnet growth as the garnet is rolled by shear along the foliation planes (the classic interpretation); (3) syntectonic garnet growth with recrystallization of the schistosity into a new orientation while the porphyroblast remains immobile (see previous page). The neutral term *rotational* is appropriate for this texture.

The matrix of the rock is composed of quartz and sodic plagioclase (twinned in some grains) with biotite defining a rather weak E–W fabric. This schistosity appears to be flattened around the garnet, suggesting that

Syntectonic porphyroblast (continued)

deformation outlasted garnet growth. Its origin is discussed further in the text (p. 95).

Locality: Connemara, Ireland. Magnification: × 20, PPL and XPL.

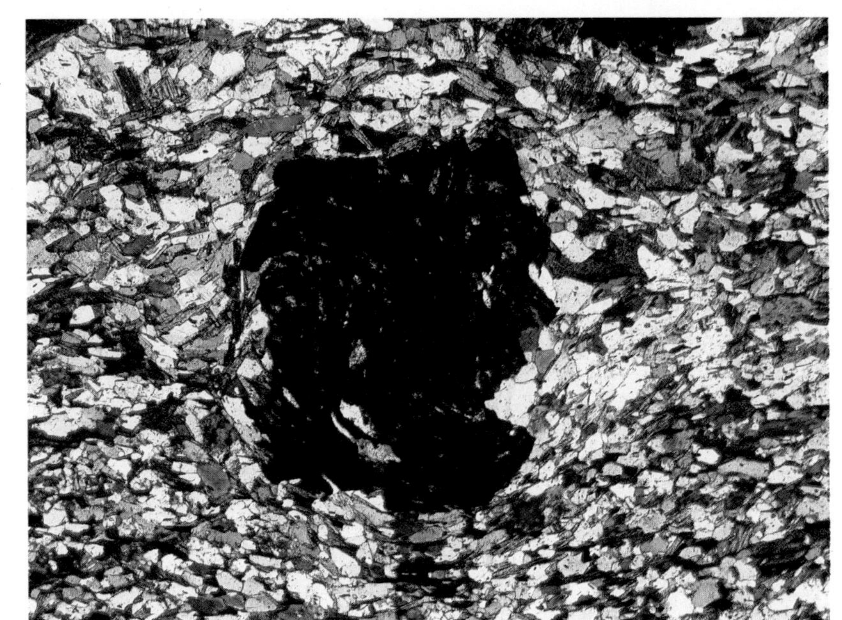

91 Late tectonic porphyroblasts in garnet muscovite schist

This rock has a very pronounced foliation due to both the segregation of the matrix minerals into quartz-rich layers (with minor plagioclase) and phyllosilicate-rich layers, and the alignment of individual phyllosilicate grains and small, tabular ilmenites. Garnet porphyroblasts overgrow this foliation and the pronounced segregation layering continues through the garnets as alternating inclusion-poor layers (corresponding to phyllosilicate-rich domains) and inclusion-rich layers where the garnet grows across a quartzose band. Minor later deformation has however led to small strains concentrated as kinks at the garnet edges. Large opaque porphyroblasts are of pyrite.

The phyllosilicate phases present are chlorite and muscovite, and the rock has undergone some retrograde metamorphism leading to growth of chlorite from garnet at its edges. Any biotite originally present must also have been chloritized.

Locality: Ben Nevis, Scotland. Magnification: × 22, PPL.

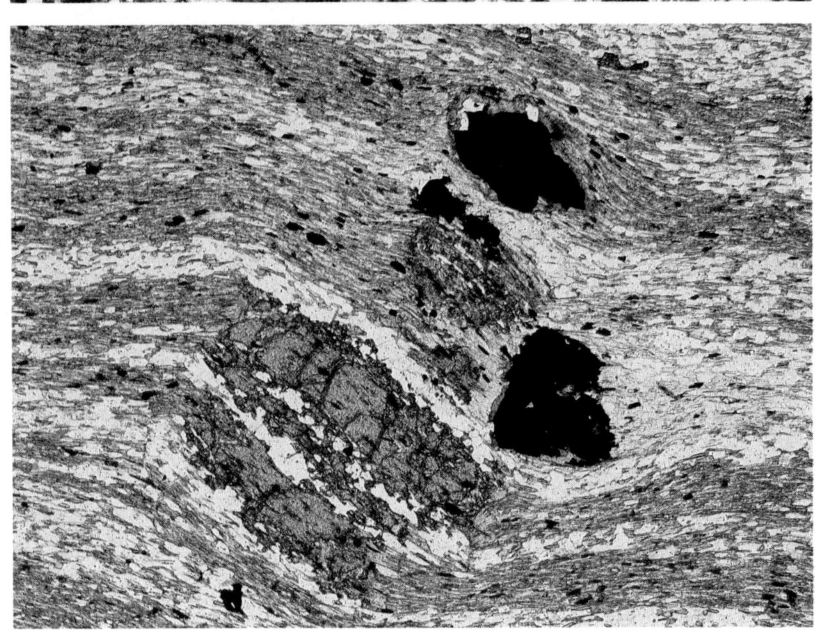

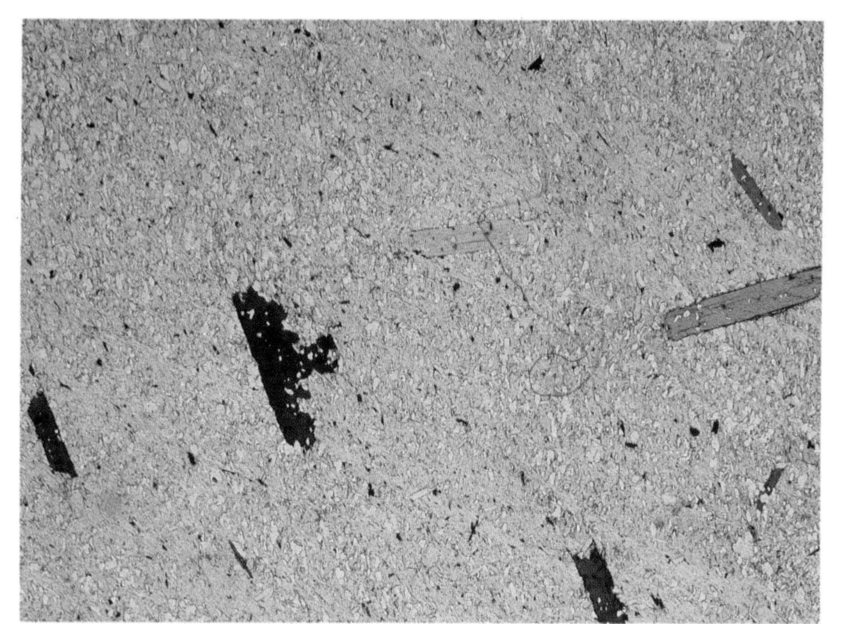

92 Post-tectonic porphyroblasts in biotite muscovite schist

This biotite zone schist reveals a complex history of deformation which preceded the growth of biotite. The rock matrix is composed largely of muscovite and quartz and has a pronounced crenulation fabric. An early, N–S fabric is still indicated by the alignment of elongate quartz grains, but these now occur in the hinges of later crenulations. Muscovite in the crenulation limbs has been re-oriented into a diagonal fabric. Randomly oriented biotites overgrow both fabrics and there is no distortion of the fabrics around the biotite porphyroblasts. Hence biotite growth was entirely post-tectonic.

Locality: Loch Leven Scotland. Magnification: × *7, PPL and XPL.*

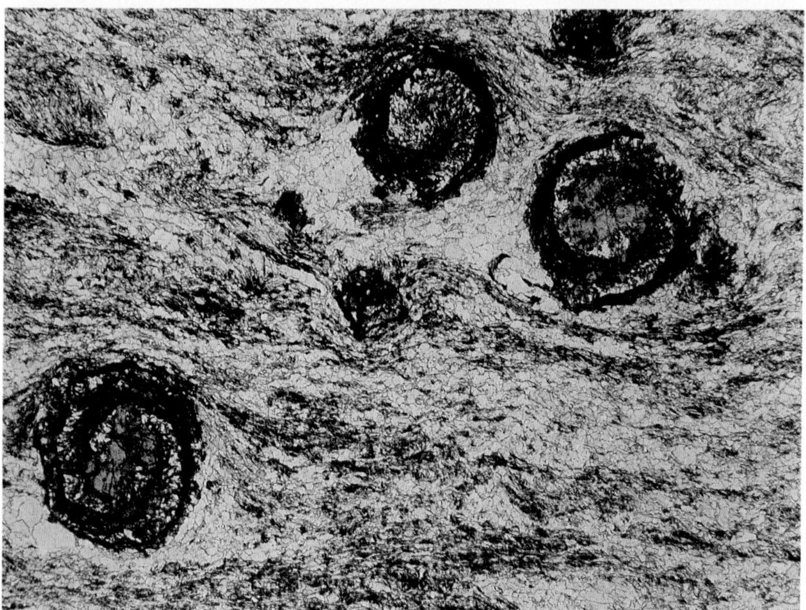

93 Syntectonic (*snowball*) porphyroblasts in calcareous schist

This is a rock of rather unusual composition, being composed predominantly of quartz and calcite with conspicuous garnet porphyroblasts. Trails of fine graphite pick out the matrix schistosity and cloud the garnets. The curved patterns of inclusions are characteristic of syntectonic porphyroblasts, and here form well-developed spirals. The higher magnification, XPL, view shows the inclusion trails in more detail and suggests a rotation of the order of 270°. It can be seen that the spiral trails are not defined by the alignment of individual quartz inclusions, but by a zone very rich in quartz, within which individual grains are aligned across the spiral, not along it. This suggests that the inclusion-rich spiral, corresponds to the hinge region of a crenulation (*see for example* **92**, *above*) developed by refolding of an earlier schistosity. This rock is illustrated further, and its origin discussed in some detail, in Rosenfeld (1968, pp. 90–1).

Snowball texture (continued)

Locality: Springfield, Vermont, USA. (Locality S41e of Rosenfeld, 1968). Magnification: × 5, PPL; and × 14, XPL.
Reference: Rosenfeld JR 1968 Garnet rotations due to the major Palaeozoic deformations in southeast Vermont. In Zen E-An, White W S, Hadley J B, Thompson J B Jr, Studies of Appalachian Geology: Northern and Maritime. Wiley pp. 185–202

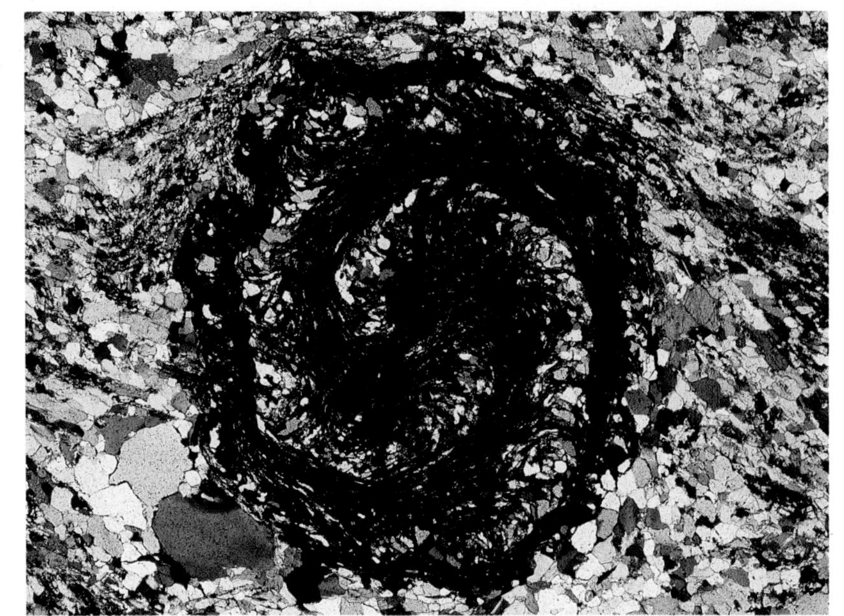

94 Multiple porphyroblast growth in staurolite garnet mica schist

Both staurolite and garnet form conspicuous porphyroblasts in this rock. Whereas staurolite forms typical pale yellow grains in the PPL image, the garnet is so heavily crowded with small cracks and/or inclusions, beyond the resolution of the petrological microscope, that it appears almost opaque. Abundant quartz inclusions in garnet define a complex internal schistosity (S_i), apparently the result of the garnet overgrowing a pre-existing crenulation fabric. The dominant schistosity of the rock (S_e), defined by muscovite and biotite, is discordant to that in the garnets and is strongly flattened around them suggesting that it formed after garnet growth. In contrast, staurolite porphyroblasts effectively overprint the schistosity which is not deformed around them, and this implies that staurolite grew later, postdating the schistosity. Nevertheless, in detail it is apparent that there is some discordance between S_i in staurolites in the central top part of the image, and S_e. This probably results from reactivation of the foliation during a still later deformation.

The dominant matrix phase, apart from micas, is quartz. Some sodic plagioclase is also present but cannot be readily distinguished in these photographs. Small tourmaline grains are present in this rock; they are zoned from green cores to yellow rims. One is quite conspicuous in the centre of the lower half of the field of view.

Locality: Connemara, Ireland. Magnification: × 11, PPL and XPL.

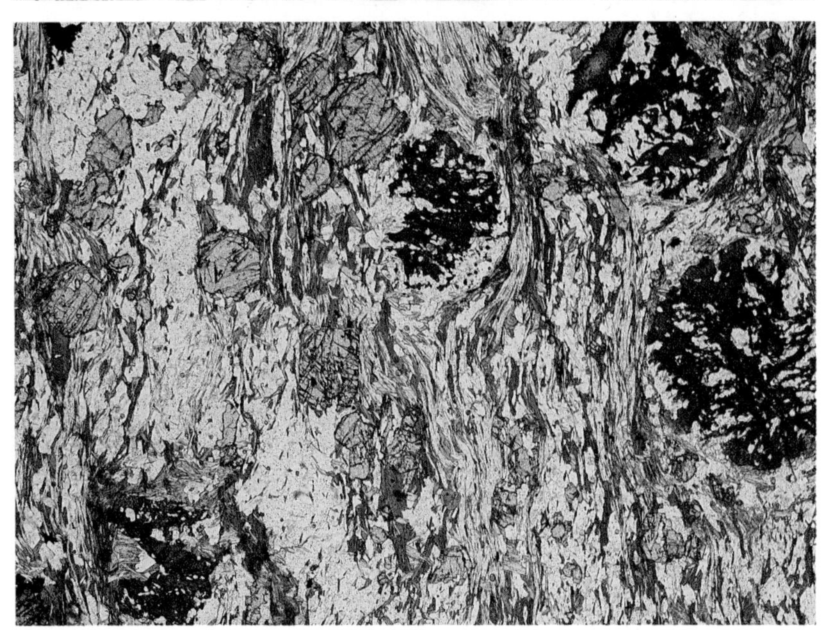

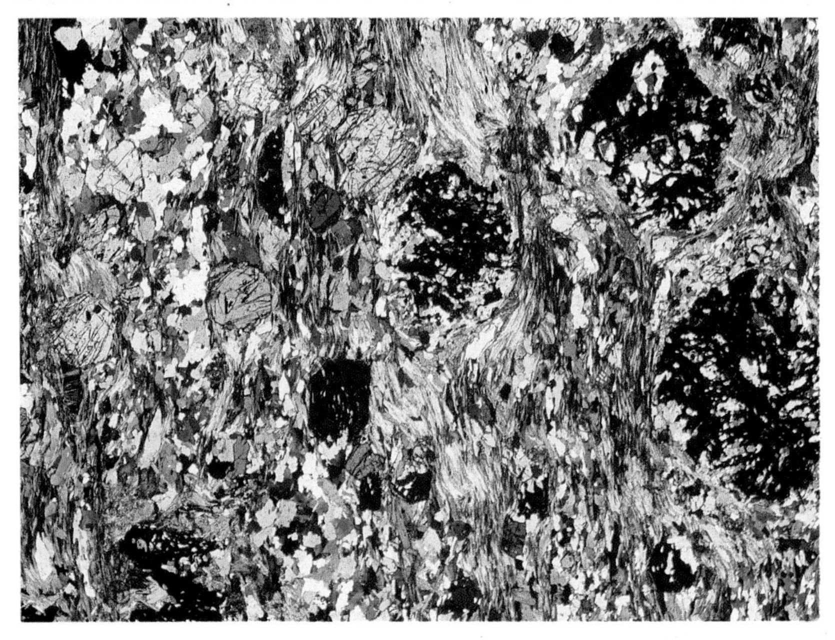

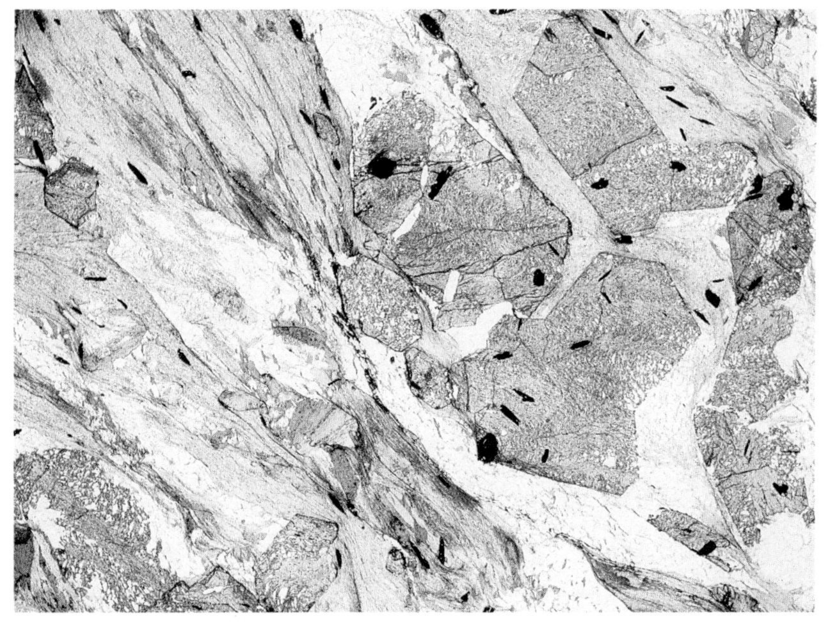

95 Complex metamorphic/deformation history in staurolite schist

The staurolite crystals in this rock contain helicitic (microfolded) inclusion trails, developed by overgrowth of an already crenulated earlier schistosity. During subsequent deformation a new matrix mica foliation has developed (aligned diagonally), but strain was heterogeneous with most intense foliation development occurring where there are fewest porphyroblasts. Note that the microfolds preserved as inclusion trails are now aligned in different orientations in each porphyroblast, because of differential rotation of staurolites during the later deformation. This is unusual, more commonly inclusion trails remain nearly parallel in all porphyroblasts after subsequent deformation (cf. **89**).

Locality: Pyrenees, Spain. Magnification: × *7, PPL and XPL.*

Reaction textures

Some metamorphic textures reflect chemical reactions that took place during metamorphism, and provide clues to the sequence of assemblages that may have been present in the rock, and hence its reaction history.

Textures which preserve relics of earlier assemblages include armoured relic inclusions (discussed above, p. **86**, e.g. **96**, **97**), zoned crystals (e.g. **98**) and reaction rims and coronas (**101–104**). In all these cases, potentially reactive minerals have become isolated from one another by a barrier of material through which diffusion is too slow to permit reaction to continue.

Pseudomorph textures also allow identification of the earlier mineralogy of a rock, though often only a distinctive grain shape allows the precursor mineral to be identified. The pseudomorphs of chiastolite (**110**) are a good example; another (not illustrated here) is the development in blueschist of intergrowths of zoisite, kyanite and quartz pseudomorphing lawsonite porphyroblasts (**66**) and formed from its high temperature breakdown near the blueschist–eclogite transition.

Pseudomorphs may also provide detailed information about the mechanism by which reactions take place. Most of the examples illustrated here involve relatively little chemical change as the pseudomorph develops, but **100** illustrates a fibrolite pseudomorph of very different composition from its garnet precursor. The development of such textures, within a framework of metamorphism that is essentially isochemical on the hand specimen scale, requires a complex network of simultaneous ionic reactions leading to complementary local chemical changes in different parts of the rock that cancel one another out overall.

96 Sillimanite garnet schist with staurolite relics

Plagioclase forms abundant porphyroblasts in this schist and is distinguished from quartz by its coarser grain size, patchy alteration, the presence of inclusions and in some instances twinning. Biotite occurs both in the matrix and as inclusions in plagioclase, and at several locations is overgrown by clusters of fibrolitic sillimanite. This is seen at two places near the left edge and near the lower edge below garnet. Garnet occurs as rather small, anhedral crystals. A large, twinned plagioclase porphyroblast in the upper right quadrant includes a cluster of heavily corroded remnants of staurolite, now isolated from the rest of the rock by the enclosing plagioclase. This is shown in more detail in the high magnification view. Staurolite is not present in the rock matrix because the peak temperature exceeded that required for staurolite to react with quartz, producing garnet and sillimanite. Staurolite inclusions in plagioclase were however isolated from quartz and were unable to participate in the reaction.

Locality: Maam Valley, Connemara, Ireland. Magnification: × 12, PPL and XPL, and × 43, PPL.
Reference: Yardley B W D, Leake B E, Farrow C M 1980 Journal of Petrology **21**: 365–99

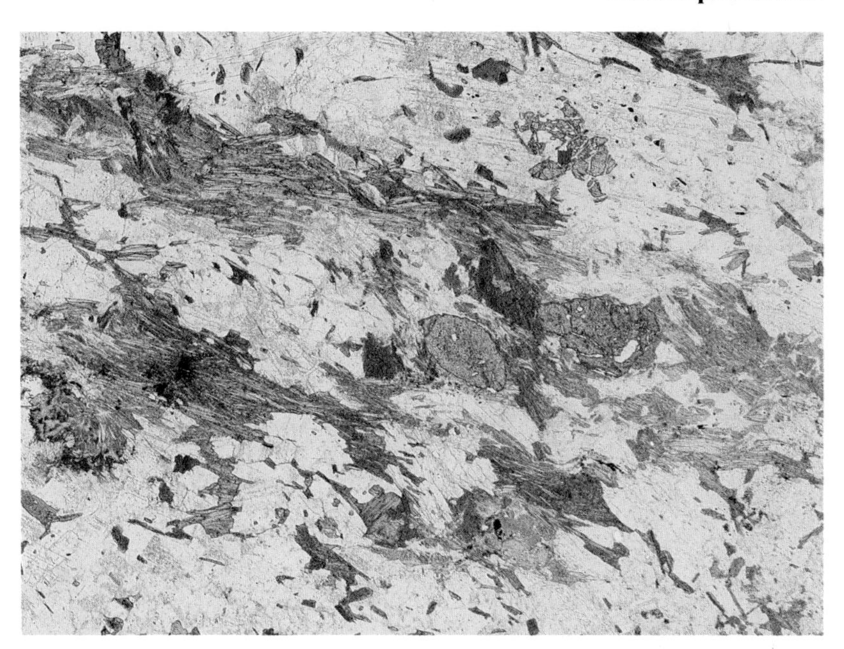

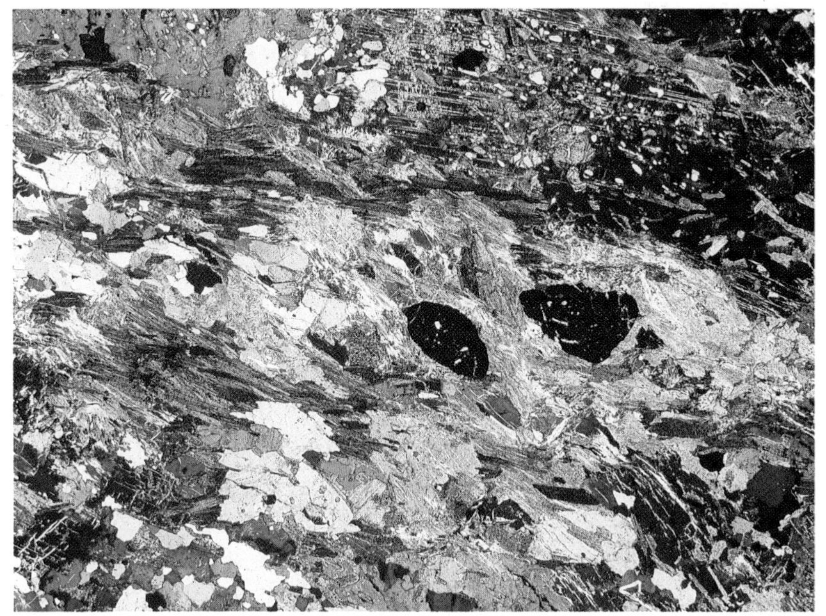

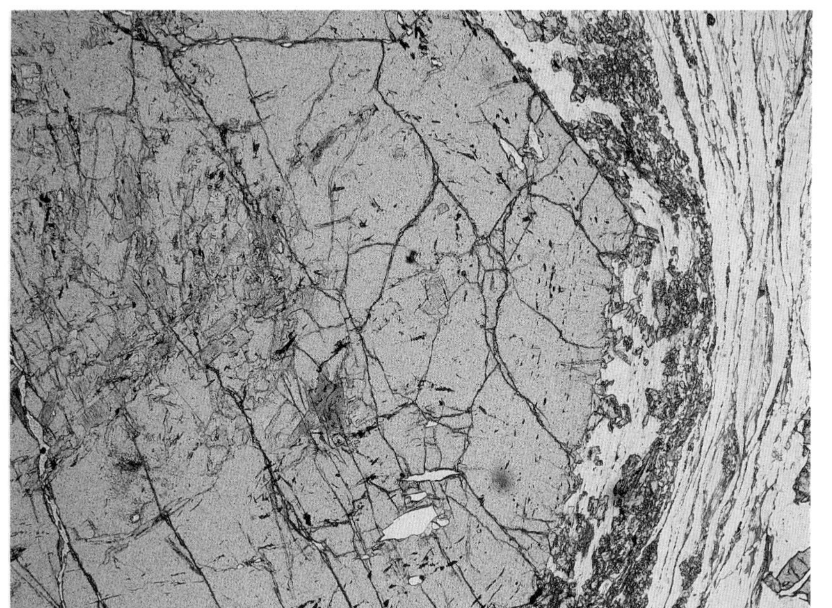

97 Staurolite garnet schist with chloritoid relics

Staurolite is a characteristic mineral of lower amphibolite facies pelites and commonly grows from the breakdown of chloritoid, a greenschist facies mineral, by reaction with quartz. In this rock large garnet porphyroblasts are set in a matrix of staurolite, muscovite, quartz, plagioclase and minor biotite. The garnet porphyroblasts contain inclusions and many of these are of pale bluish-green chloritoid, even though no chloritoid is present in the matrix. These inclusions are known as armoured relics (or simply relics) because they are of a mineral which has broken down in the rock matrix due to changes in P and T, but survives inside garnet because, being entirely surrounded by garnet, it is completely isolated from quartz with which it would otherwise react. While the reaction of chloritoid with quartz to form staurolite marks the beginning of the amphibolite facies, chloritoid in isolation is stable to appreciably higher temperatures.

Locality: Zwenbergertal, near Obervellach, Austria. Magnification: × 15, PPL.

98 Zoned crystals in meta-ironstone

This rock is an unusually iron-rich amphibolite, composed predominantly of quartz, magnetite and amphibole. Small amphibole grains in the matrix are pleochroic between deep blue, lilac and a pale green, and are sodic amphiboles ranging from riebeckite to crossite. The rims of two large amphibole grains are of similar composition to the matrix grains, but their interiors are strongly zoned. The cores of these amphibole porphyroblasts are of pale cummingtonite, in marked contrast to the strongly pleochroic sodic rims. There is also patchy development of green ferroactinolite within some of the zoned grains. Only the rims of these amphiboles can be considered as in equilibrium with the rest of the minerals in the rock; the cores preserve compositions formed earlier in the rock's history under different conditions which have survived because of the sluggishness of volume diffusions through amphibole.

In addition to the major constituents, this rock contains a few small green grains of aegirine-augite associ-

Zoned crystals (continued)

ated with the matrix amphibole. Some of the smaller black grains in the matrix are of deerite, which has an acicular habit with diamond shaped cross section, and is not quite truly opaque. Garnet occurs elsewhere in the thin section, but is not shown here.

Locality: Sifnos, Greece. Magnification: × 16, PPL and XPL.
Reference: Evans B W 1984 Geological Society of America: Abstracts with Programs **16**: 504

99 Atoll structure in garnet mica schist

The garnet texture shown here is one whose origin has been controversial for many years. Several of the garnet grains illustrated contain greenish-brown biotite, sometimes with muscovite also, in their cores. These mica grains are at least as large as similar mica grains in the matrix and appear to be replacements of the garnet core rather than inclusions trapped during garnet growth. The resulting atoll texture comprises a shell of garnet with near euhedral outline but irregular inner edges and a filling of other phases. The rock shows strong segregation into an upper micaceous portion and a lower band rich in quartz and plagioclase. There are a number of small green grains of accessory tourmaline in the micaceous layer.

Locality: Meall Druidhe, Kinloch Rannoch, Scotland. Magnification: × 34, PPL and XPL.

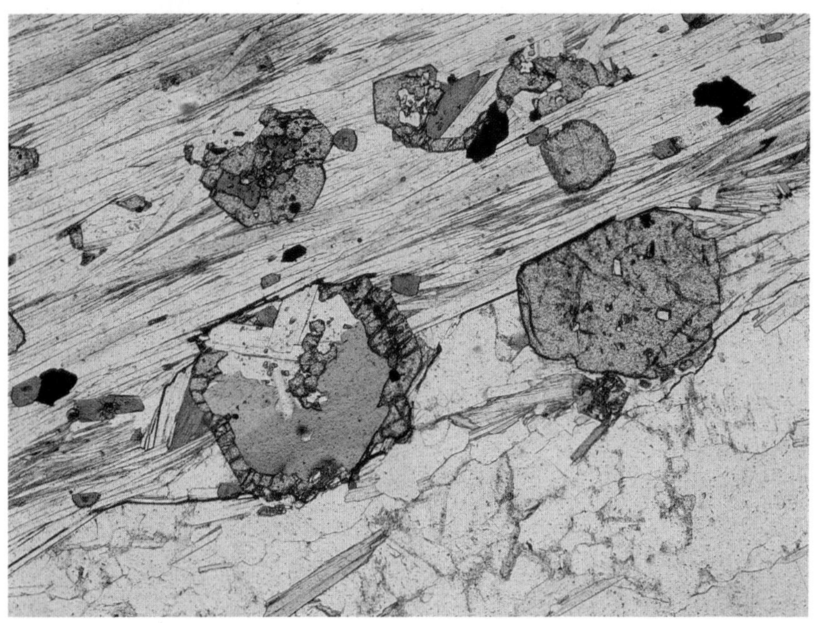

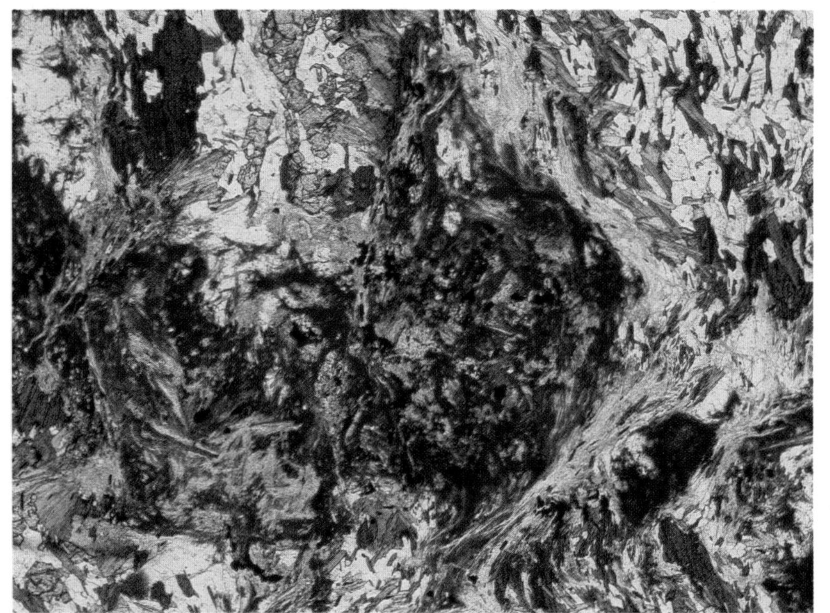

100 Pseudomorph textures in staurolite sillimanite schist

Fibrolitic sillimanite often occurs in clusters or segregations, and where the fibrolite needles are small, such clusters can appear almost opaque in PPL. This is because there is usually interstitial quartz and the strong contrast in relief between the two phases, repeated many times in a small volume, produces an opaque effect. In this view, a fibrolite-rich cluster has the morphology of pre-existing garnet, which it replaces; some other segregations in the rock retain remnants of garnet. Other phases in the rock include yellow, high relief staurolite and biotite; quartz and plagioclase form the low relief, colourless matrix. There is a trace of green tourmaline. The truly opaque phase is ilmenite, and quite large ilmenite grains occur with fibrolite in the garnet pseudomorph.

The formation of this type of pseudomorph requires considerable local mass transfer; it is for example much richer in Al and poorer in Fe than the original garnet. Overall, however, the rock composition is unchanged; garnet, staurolite and muscovite have reacted to form sillimanite and biotite. A complex cycle of local ionic reactions (Carmichael, 1969) has permitted complementary chemical changes in different parts of the rock. A less advanced stage of the replacement of garnet in this rock suite is illustrated in **19** (*see* Yardley, 1977).

Locality: Maam Valley, Connemara, Ireland. Magnification: × *16, PPL and XPL.*

References: Carmichael D M 1969 Contributions to Mineralogy and Petrology **20**: 244–67
Yardley B W D 1977 Contributions to Mineralogy and Petrology **65**: 53–8

101 Reaction rims in sapphirine granulite

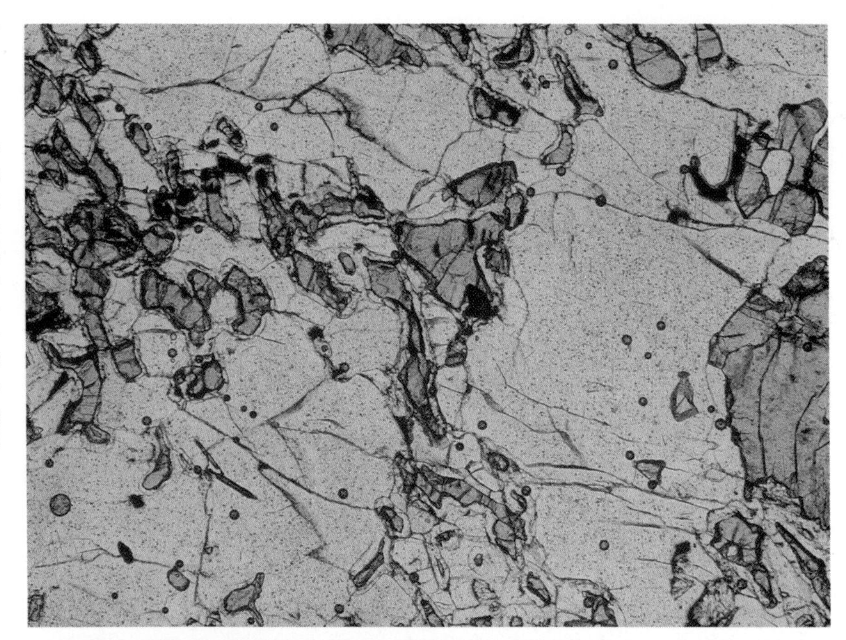

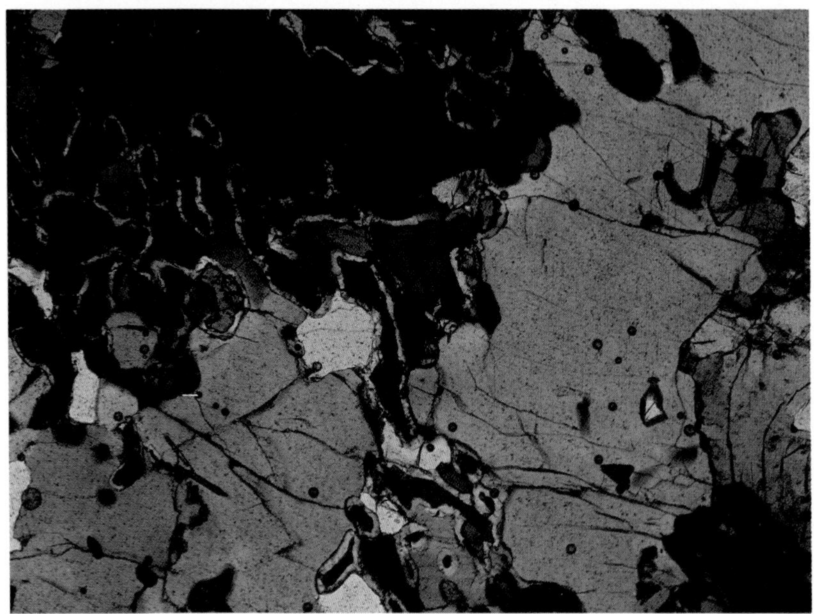

Under very high temperature conditions, in the upper part of the granulite facies, sapphirine ($(Mg,Fe)_2Al_4SiO_{10}$) can coexist with quartz. This rock contains rounded blebs of blue sapphirine (low birefringence) in a matrix of low relief, coarse-grained quartz. A few grains of pale brown hypersthene are also present. Sapphirine grains are mantled by a thin colourless halo of cordierite (also low relief and low birefringence). These reaction rims of cordierite are a retrograde phenomenon, probably produced by the reaction of sapphirine with orthopyroxene, taking place during cooling and uplift. Note that there are some conspicuous small air bubbles present.

Locality: Crosby Nunataks, Enderby Land, Antarctica.
Magnification: × 38, PPL and XPL.

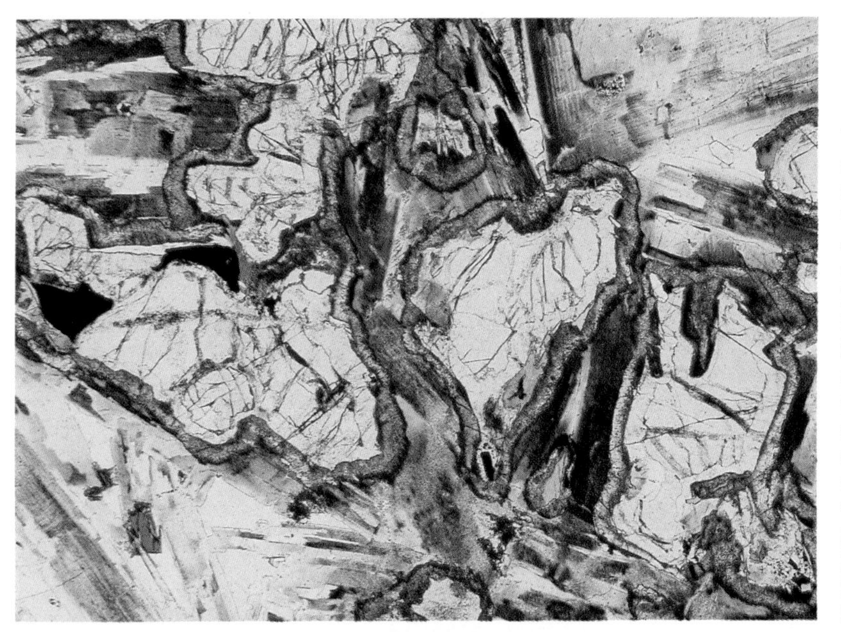

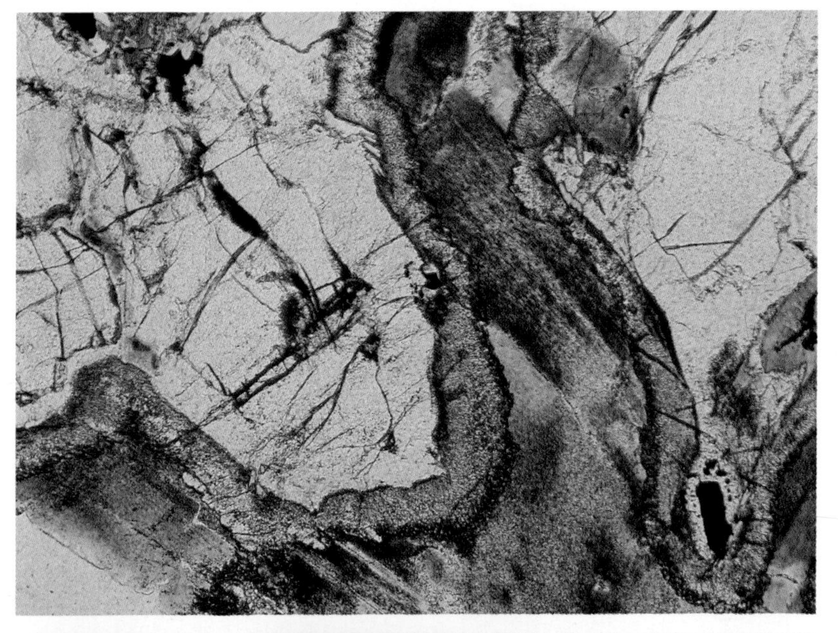

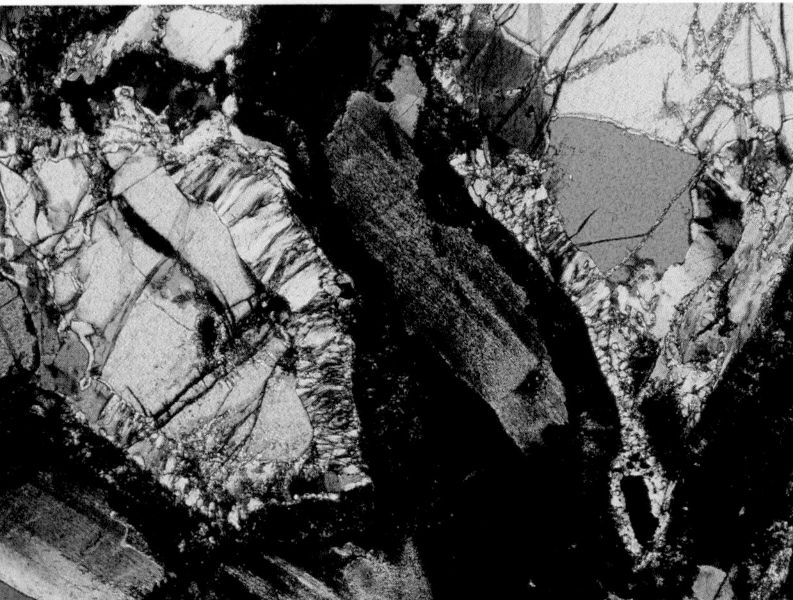

102 Corona texture I: in metamorphosed olivine dolerite

This rock retains a coarse igneous texture typical of dolerite (diabase) and is virtually undeformed despite extensive metamorphic reaction. The higher power views show an enlargement of the area below the centre of the field of view above. Primary plagioclase, which is unstable with olivine under the conditions of the granulite facies metamorphic event, has taken on a patchy brown colouration in PPL which is caused by fine scale inclusions but serves to pick out original igneous grain shapes and twinning. Original igneous olivine grains are now mantled by complex coronas where they were in contact with the plagioclase. In PPL it is the high relief outer zone of the corona that is clearly apparent. This appears almost black in XPL and is composed of a very finely intergrown symplectite of clinopyroxene and plagioclase with an outer fringe of fine garnet. In XPL the olivine grains are seen to be mantled by an inner corona of orthopyroxene with fibrous habit, and indeed the small contrast in relief between olivine and orthopyroxene can be detected in the PPL view. Small amounts of brown biotite occur locally within the coronas, usually at the interface between the orthopyroxene and symplectite zones.

Locality: Midøy, west Norway. Magnification: × 27, PPL; and × 72, PPL and XPL.

103 Corona texture II: in metamorphosed dolerite

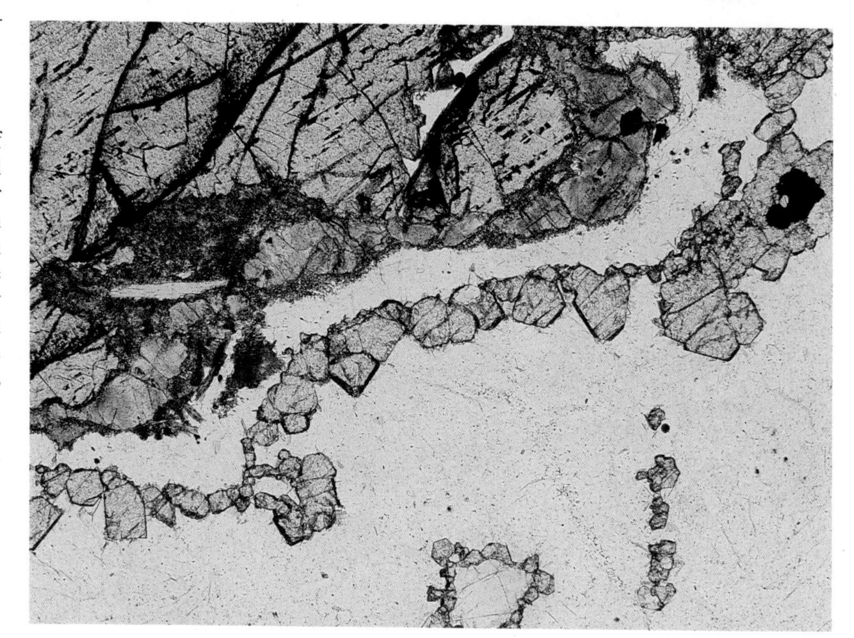

This photograph shows a slightly different type of corona, developed at the interface between original igneous augite (upper left) and plagioclase. The inner zone of the corona, in contact with augite, is of pale green omphacitic clinopyroxene. A fine fringe of deeper green retrograde amphibole often surrounds the omphacite grains, and biotite is also patchily developed, notably towards the left side. The omphacite zone is separated by a clear, low relief zone of recrystallized plagioclase from an outer zone of small garnet crystals, which marks the outer limit of the corona texture.

Locality: Fiskå, Sunnmøre, west Norway. Magnification: × 22, PPL.

104 Corona texture III: in metamorphosed dolerite

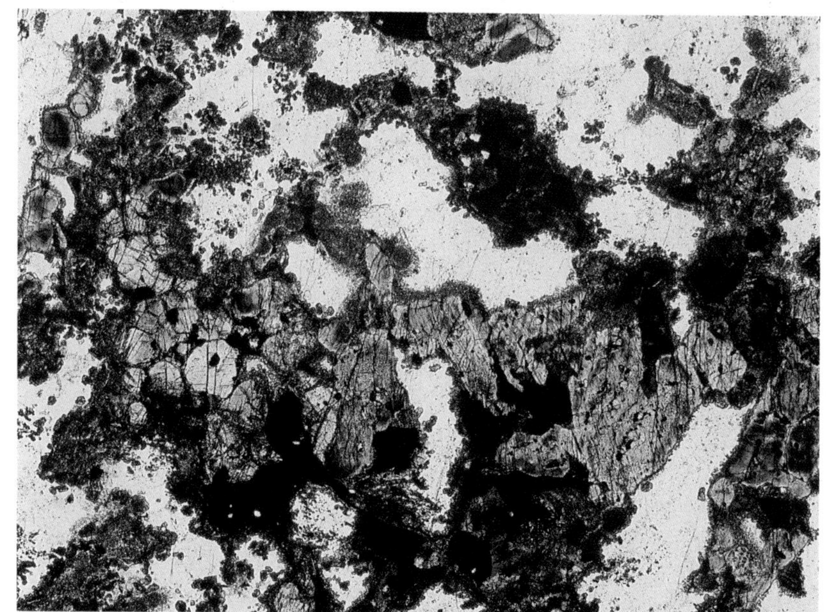

In this example the development of coronas at the interface between original igneous pyroxene and plagioclase has been accompanied by more thorough recrystallization. Conspicuous rims of fine garnets separate plagioclase domains from pyroxene domains but the original igneous phases are extensively recrystallized. Plagioclase has been replaced by albite with complex sub-grain structures, while augite is largely pseudomorphed by either green omphacite or pale brown orthopyroxene. In addition, strongly coloured brown biotite and green hornblende are present, notably in the lower half of the field of view, and are probably of retrograde origin. An opaque phase is also conspicuous.

Locality: Hellesylt, Sunnmøre, west Norway. Magnification: × 16, PPL and XPL.

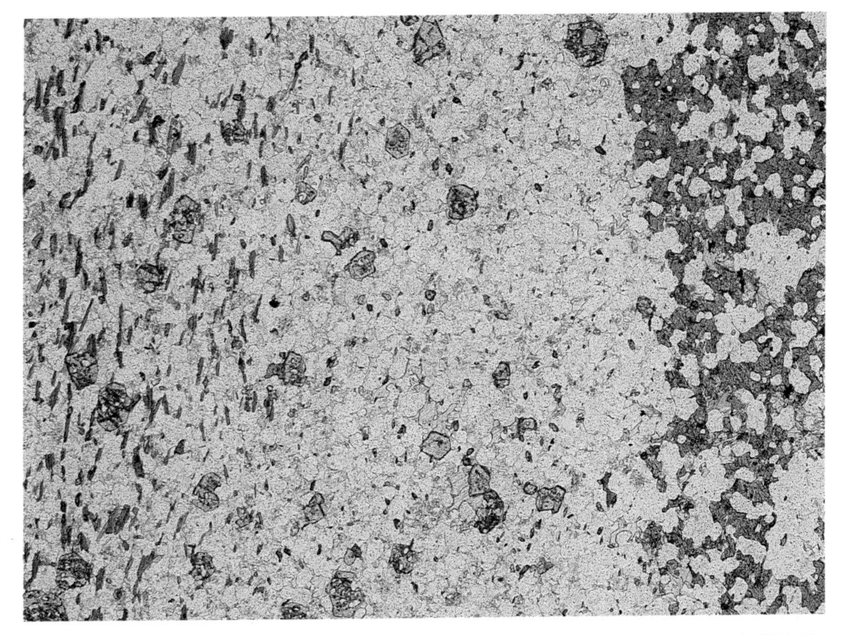

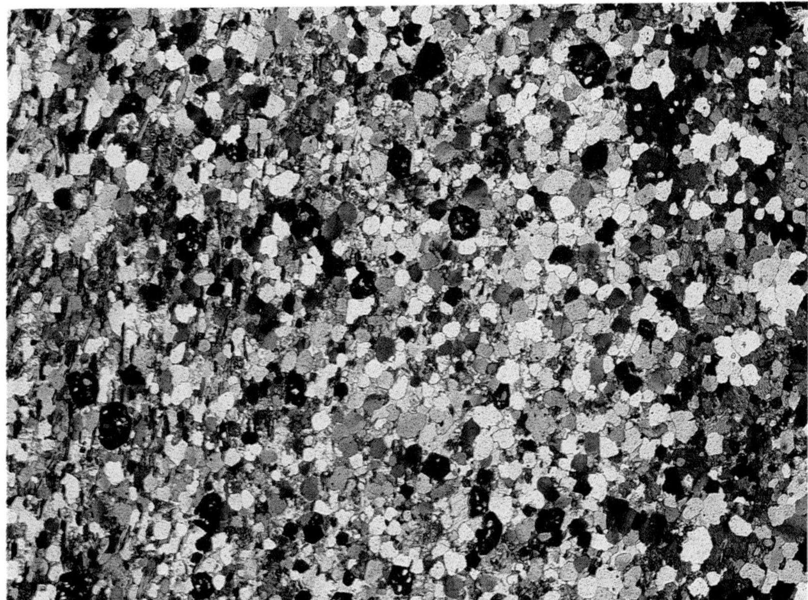

105 Diffusion metasomatic zoning in calc-silicate rock

Sedimentary processes can juxtapose layers of contrasting mineralogy which subsequently react together during medium to high grade metamorphism. Such reaction produces a zone, or zones, of product minerals along the interface whose width is constrained by the ease with which material can diffuse into the reaction zone from the unaltered beds on either side. The chemical change taking place between the two original layers is termed diffusion metasomatism, and the most widespread example is the formation of layers of calc-silicate minerals at the interface between marble and schist. Thin marble layers may be destroyed entirely, leaving a calc-silicate layer only. This view shows the interface between schist on the left and calc-silicate on the right; probably originally limestone. The schist layer is composed of biotite, plagioclase and quartz with garnet; biotite becomes sparser towards the calc-silicate layers. The central zone is of quartz, plagioclase and garnet with minor calcite (for example around garnet in the lower part of the field of view) and small lozenge-shaped sphenes. Feldspar is distinguished from quartz by a slightly clouded appearance in PPL, and occasional twinning. The right hand, calc-silicate, zone is predominantly of hornblende and quartz with minor biotite and calcite.

Locality: Loch Assapol, Ross of Mull, Scotland. Magnification: × 11, PPL and XPL.

Textures of polymetamorphism

It is not uncommon for metamorphic rocks to contain minerals formed at different times and under different physical conditions. Indeed, the most difficult part of many metamorphic studies is often determining which minerals in a rock actually coexisted together at equilibrium (Yardley, 1989; pp. 46–49).

The superimposition of different metamorphic events on a single suite of rocks is known as *polymetamorphism*, and this term is usually used where the different events recorded by the rocks are not both part of a single cycle of heating and cooling. Where the second metamorphic event takes place at a higher temperature than the first, the original assemblage usually reacts completely, as in prograde metamorphism, although textural relics may remain. An example is the slaty matrix fabric apparent in **1**. Only where driving forces are very small, as for some polymorphic transitions, do lower temperature minerals survive (e.g. **111**). On the other hand if the second event is a lower temperature one (i.e. retrograde), many of the reactions that might occur require the reintroduction of fluid to the rock, and if the supply of fluid is limited the reactions cannot go to completion and minerals from both the primary and retrograde event will be present in the final rock. In addition to temperature differences, there may also be appreciable pressure differences between the events to which a polymetamorphic rock has been subjected.

Of the examples illustrated here, **107**, **108** and **109**, each illustrate different types of overprinting in which water must be added in order for the assemblages characteristic of the second event to develop. Effects of a later, lower pressure, contact event on higher pressure regional assemblages are shown in **106** and **112**. **113** also illustrates the effect of a drop in pressure, though it is possible that this is not strictly a polymetamorphic rock. Likewise **111** is not strictly polymetamorphic, but is included here with other examples of polymorphic transitions.

106 Contact metamorphism after regional in andalusite garnet schist

An originally euhedral garnet formed during regional metamorphism has here been extensively broken down during a subsequent contact event. The outer parts of the garnet (where original shape is best seen in XPL) are replaced by quartz, plagioclase, muscovite, green biotite and magnetite, while much of the core remains intact. Magnetite tends to occur in discrete planes that probably mark progressive replacement along cracks in the early stages of garnet breakdown. The rock matrix is dominated by muscovite, often recrystallized to a decussate texture and there are large andalusite porphyroblasts formed during the contact event in the corners of the field of view. The garnet breakdown reaction here reflects oxidation as well as changes in P and T. It was approximately: garnet + 0_2 = plagioclase + magnetite + andalusite + quartz.

Locality: Easky Lough, Co Mayo, Ireland. Magnification: × 8, PPL and XPL.
Reference: Yardley B W D, Long C B 1981 Mineralogical Magazine **44**: 125–31

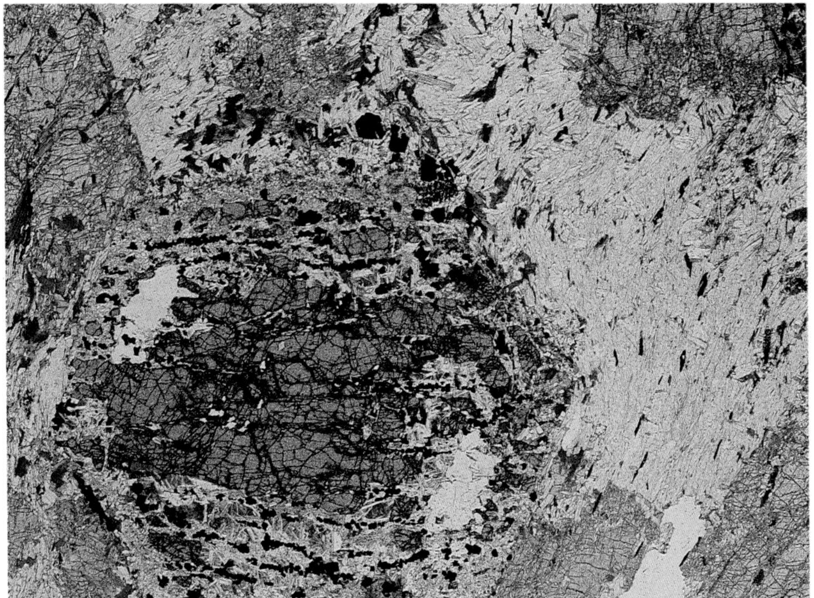

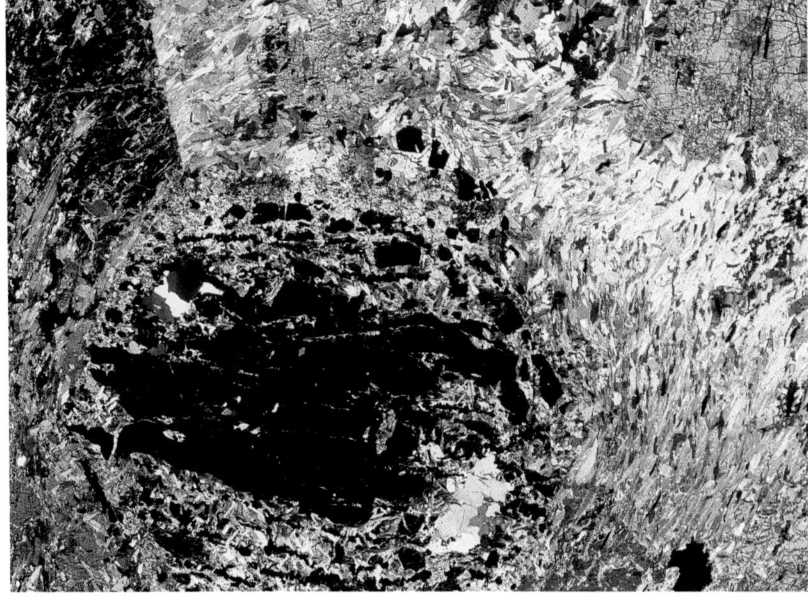

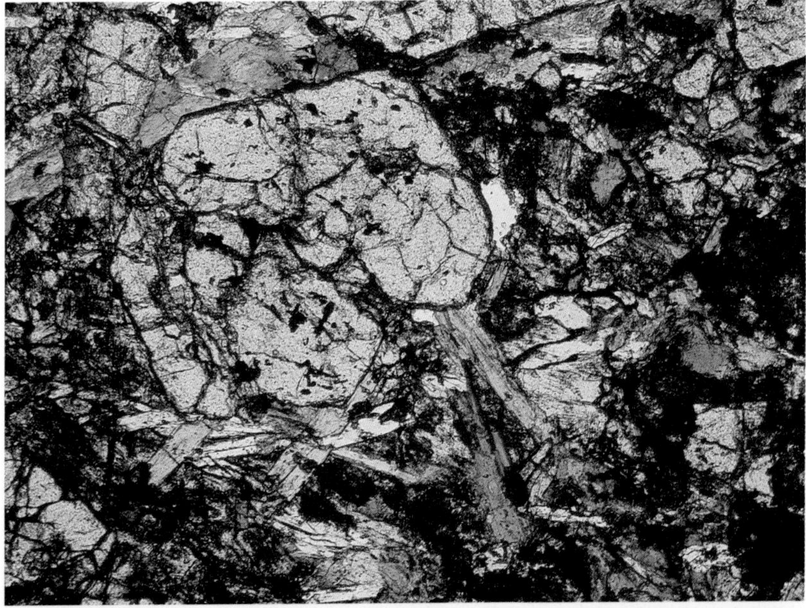

107 Blueschist overprint of eclogite

The thin section illustrated here has been made slightly thick in order to emphasize the body colours of the minerals. The mineralogy of this rock is intermediate between the blueschist facies and the eclogite facies, since garnet, glaucophane and omphacite are all present. Dark, nearly opaque, clouded areas are of rutile mantled by sphene. The lower half of the low power view is essentially eclogite, dominated by garnet and green omphacite with accessory sphene, but glaucophane is also present. The enlarged views are from the lower right quadrant and show the spectacular blue-lilac-colourless pleochroism of the glaucophane. In the upper half of the low power view, omphacite has been largely replaced by glaucophane, and some chlorite is present.

Locality: Franciscan Formation, Jenner, California, USA. Magnification: × 7, PPL; and × 25, PPL and XPL.

108 Greenschist facies overprint on blueschist (retrogressed blueschist)

This metabasite has a pronounced schistosity defined by the alignment of pale glaucophane crystals and contains corroded relics of garnet that probably also formed under high pressure conditions. Green chlorite has grown extensively at the expense of garnet, and also occurs with muscovite in pressure shadows by garnet. Glaucophane has partially broken down to albite and tremolite–actinolite and the resulting intergrowths of aligned amphibole in porphyroblastic albite are illustrated in the enlarged view which shows the lower left quadrant of the lower power image. The opaque grains are of rutile, mantled by sphene.

Locality: Val Chiusella, Italy. Magnification: × 25, PPL and XPL; and × 72, XPL.

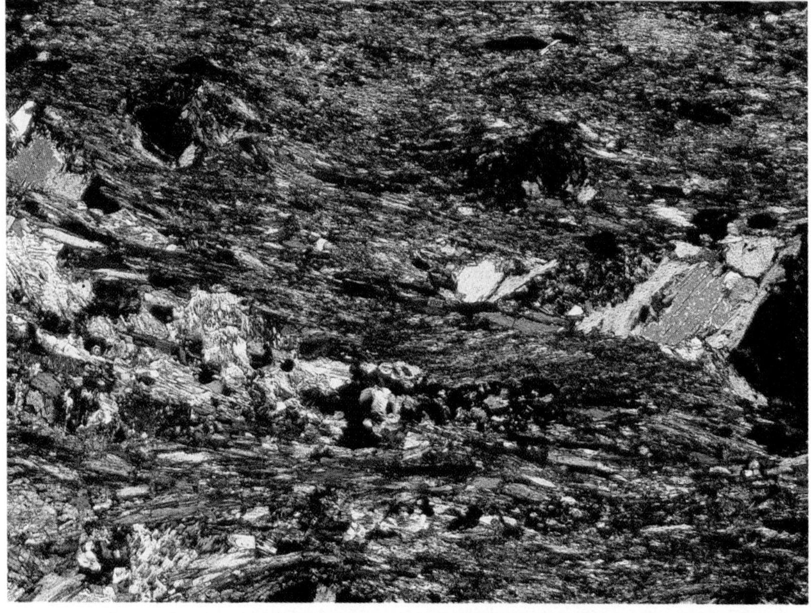

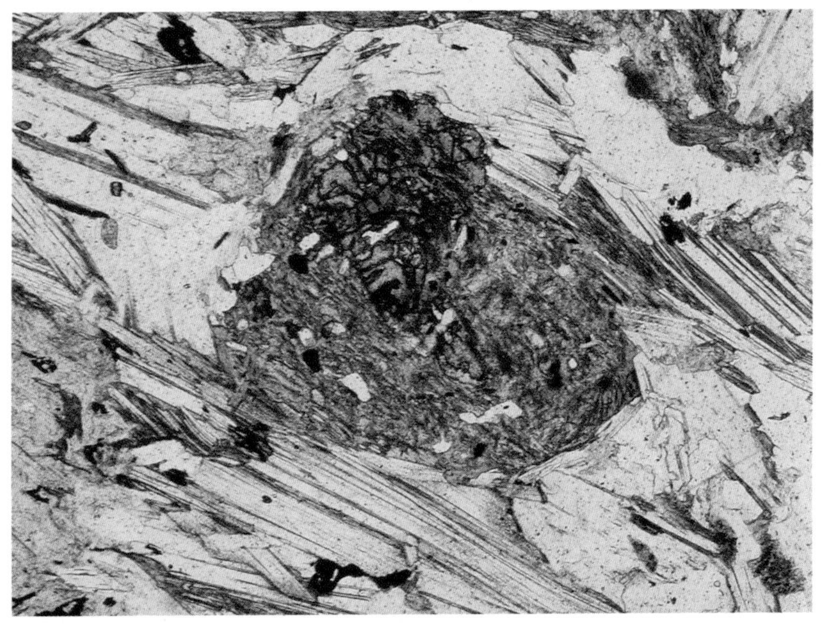

109 Retrogressed garnet mica schist

In the centre of field of view is a pseudomorph after garnet, composed predominantly of chlorite but with a large grain of chloritoid replacing the upper corner (higher relief, blue-green in PPL). Some relics of the original garnet remain, and a number of quartz grains within the pseudomorph probably represent inclusions in the original garnet. Biotite has been heavily chloritized and now has a patchy greenish-brown colour, matched by patchy birefringence in XPL. The XPL image shows several areas of shimmer aggregate i.e. patches of fine-grained white micas with bright birefringence colours. Although muscovite usually predominates, paragonite may be present. These patches of shimmer aggregate are likely to be pseudomorphs after a high temperature phase, either staurolite or kyanite, but only where relics remain can an unequivocal identification be made.

The occurrence of chloritoid as a retrograde mineral after garnet shows that the rock suffered a lower greenschist facies retrograde overprint on an original amphibolite facies assemblage.

Locality: Rosses Point, Co Sligo, Ireland. Magnification: × 27, PPL and XPL.
Reference: Yardley B W D, Baltatzis E B 1985 Contributions to Mineralogy and Petrology **89**: 59–68

110 Polymorphic transition of kyanite after chiastolite in graphitic hornfels

The clear areas in the centre and left of the field of view define the shape of a prismatic andalusite crystal, and the cross shape, with concentrations of graphite at some faces, suggests that the parental crystal was of the chiastolite variety (cf. **27**). No andalusite remains in this slide; the low relief material in the pseudomorph is a fine-grained shimmer aggregate of muscovite, and this contains radiating bundles of high relief kyanite. In addition to graphite, the matrix contains biotite and quartz with minor muscovite.

Locality: south of Dusky Sound, Fiordland, New Zealand. Magnification: × 11, PPL and XPL.

Polymorphic transition (continued)

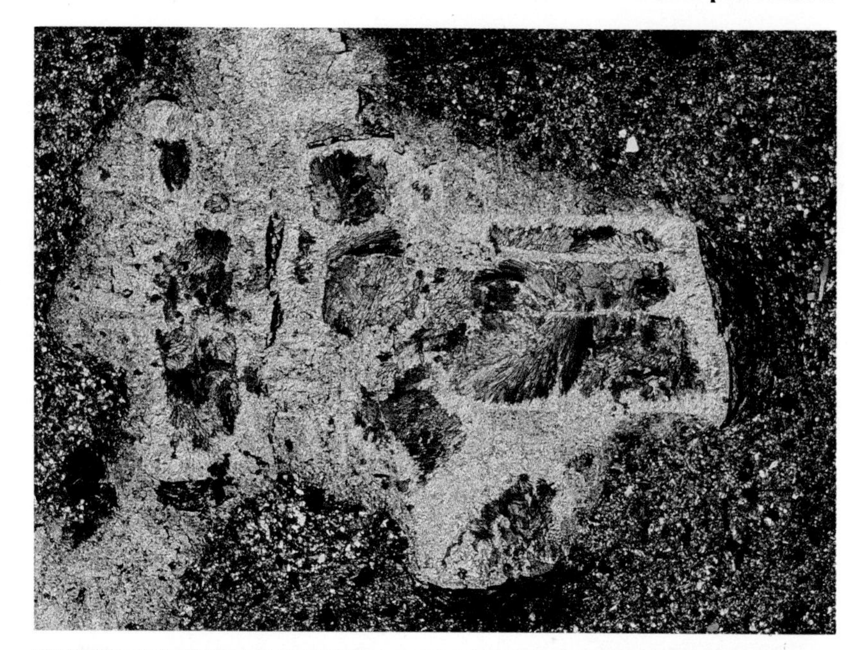

111 Polymorphic transition of sillimanite after andalusite in sillimanite hornfels

Except in the granulite facies, prismatic sillimanite usually forms only as a replacement of pre-existing andalusite porphyroblasts, and this is demonstrably the case in the rock illustrated. The field of view is dominated by a large sillimanite grain with high relief and blue birefringence colour. To both right and left this grain is flanked by lower relief andalusite with yellow birefringence, and some stripes of andalusite survive enclosed by sillimanite. This is an example of topotactic replacement of andalusite by sillimanite, and it is notable that the growth style of sillimanite is different beyond the limits of the original andalusite. At its upper termination, the sillimanite passes into distinct fine parallel fibres, while small bundles of random fibrolite have developed in biotite below the left margin. The matrix phases in this rock are quartz (with minor plagioclase), biotite, minor muscovite and an opaque phase.

Locality: Mount Stuart, northern Cascades, Washington, USA. Magnification: × 14, PPL and XPL.

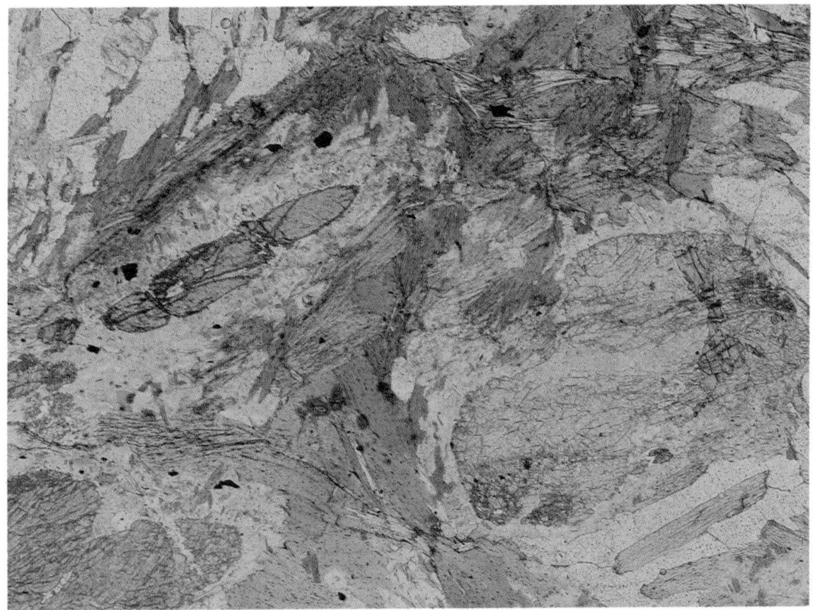

112 Polymorphic transition of andalusite after kyanite in kyanite mica schist following contact metamorphism of schist

The field of view illustrates three distinct A1-silicate grains, each mantled by a fine-grained shimmer aggregate of white mica with biotite, in a matrix dominated by biotite and quartz with plagioclase. The upper left porphyroblast is a rounded remnant of kyanite and has distinctly higher relief than the superficially similar porphyroblast in the lower left corner, which is of andalusite. The large porphyroblast on the right is composed predominantly of andalusite, with a complex texture of sub-grains, but contains a distinctly higher relief remnant of kyanite within it. A small amount of cordierite is present in this rock but is not readily visible here. A small vein of pinite cuts across the andalusite grain in the lower left corner.

Locality: Ardanalish Bay, Ross of Mull, Scotland. Magnification: × 16, PPL and XPL.

113 Polymorphic transition showing topotactic replacement of aragonite by calcite

The large carbonate grain occupying most of the field of view is of aragonite, and formed in a vein during high pressure, low temperature metamorphism. The aragonite is partially replaced by calcite which has nucleated at numerous sites around the edge of the aragonite and within it along cracks. Note that there are two contrasting morphologies of calcite. Some occurs as texturally mature granoblastic aggregates (e.g. along the lower edge of the main aragonite grain, and at the top), while elsewhere calcite forms diffuse dendritic grains within aragonite.

Locality: Eel River, north California, USA. Magnification: × 24, PPL and XPL.
Reference: Carlson W D, Rosenfeld J L 1981 Journal of Geology **89**: 615–38

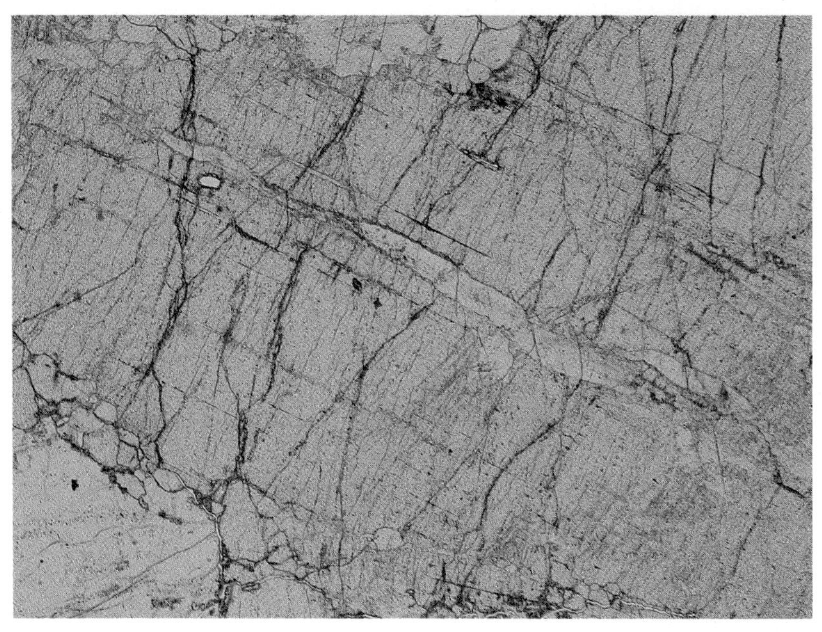

References

Barker A J 1990 *Metamorphic Textures and Microstructures* Blackie, Glasgow

Barrow G 1983 On an intrusion of muscovite biotite gneiss in the S E Highlands of Scotland and its accompanying metamorphism. *Quarterly Journal of the Geological Society London* **49**: 330–58

Bell T H 1981 Foliation development: the contribution, geometry and significance of progressive bulk inhomogeneous shortening. *Tectonophysics* **75**: 273–96

Bell T H 1985 Deformation partitioning and porphyroblast rotation in metamorphic rocks: a radical reinterpretation. *Journal of Metamorphic Geology* **3**: 109–18

Bell T H, Johnson S E 1989 Porphyroblast inclusion trails: the key to orogenesis. *Journal of Metamorphic Geology* **7**: 279–310

Coombs D S 1954 The nature and alteration of some Triassic sediments from Southland, New Zealand. *Transactions of the Royal Society of New Zealand* **82**: 65–109

Eskola P 1915 On the relations between the chemical and mineralogical composition in the metamorphic rocks of the Orijarvi region. *Bulletin de la Commission Geologique de Finland,* 44

Fyson W K 1975 Fabrics and deformation of Archean metasedimentary rocks, Ross Lake – Gordon Lake area, Slave Province, Northwest Territories. *Canadian Journal of Earth Sciences* **12**: 765–76

Fyson W K 1980 Fabrics and emplacement of an Archean granitoid pluton, Cleft Lake, Northwest Territories. *Canadian Journal of Earth Sciences* **17**: 325–32

Ramsay J G 1962 The geometry and mechanics of formation of 'similar' type folds. *Journal of Geology* **70**: 309–27

Rast N 1958 The metamorphic history of the Schiehallion Complex. *Transactions Royal Society of Edinburgh* **63**: 413–31

Rosenfeld J L 1970 Rotated garnets in metamorphic rocks. *Geological Society of America Special Paper*, 129

Spry A 1969 *Metamorphic Textures* Pergamon, Oxford

Vernon R H 1975 *Metamorphic Processes* Halsted, New York

Vernon R H 1989 Porphyroblasts–matrix microstructural relationships: recent approaches and problems. *In* Daly J S, Cliff R A, Yardley B W D (eds) *Evolution of Metamorphic Belts. Geological Society of London Special Publication*, 43 pp. 83–102

Voll G 1960 New work on petrofabrics. *Liverpool and Manchester Geological Journal* **2**: 503–67

Winkler H G F 1976 *Petrogenesis of Metamorphic Rocks* 4th edn. Springer-Verlag, New York

de Wit M J 1976 Metamorphic textures and deformation: a new mechanism for the development of syntectonic porphyroblasts and its implications for interpreting timing relationships in metamorphic rocks. *Geological Journal* **11**: 71–100

Yardley B W D 1989 *An Introduction to Metamorphic Petrology* Longman, Harlow

Zwart H J 1962 On the determination of polymetamorphic mineral associations, and its applications to the Bosost area (central Pyrenees). *Geologisch Rundschau* **52**: 38–65

Index of mineral names

Note: In the case of the most common minerals, only the better examples are listed. All numbers refer to pages.

General index

Page numbers in *italics* are to entries in the general text, other page numbers refer to photomicrographs and their descriptions.